taking flight

Guilhem Lesaffre

HACHETTE ILLUSTRATED UK

contents

Autumn

The irresistible call of the south   20

And still they leave...   32

Fully equipped for flight   38

journey to the end of the night   44

skyways   52

ports of call   58

vagrants   74

The majestic flight of the crane   78

winter

short-haul flyers   88

winter quarters   96

Fleeing bad weather   106

mountain migrants   112

invaders   116

spring

The flight home   124

display time   138

Happiness is a nest   152

The young take flight   166

departure   170

practical hints   180

TWICE A YEAR, THE SKIES OF EUROPE are filled with billions of birds, day and night. Some of these migrants are content to travel modest distances within the borders of a single continent, while others continue their voyage as far as Africa or even to the edges of the Antarctic ice.

preface

Whether they weigh just $\frac{1}{3}$ ounce or have a wingspan of $6\frac{1}{2}$ feet, they are capable of covering thousands of miles, arousing astonishment, even fascination, in us poor grounded bipeds. The mystery of the irresistible urge that drives these migrations back and forth partly explains the attraction they have always held for human beings. From Japanese artists to Audubon, from Beethoven to Olivier Messiaen, from Masai warriors to Shaolin monks, and from Selma Lagerlöf to Ivan Turgenev, painters, engravers, composers, dancers, writers, and poets in civilisations far apart both in time and in space have celebrated bird migration in their works. These great journeys have even taken on a religious significance, as they did for the Romans and North American Indians.

Today, migration chiefly interests amateur ornithologists and researchers who are trying to find out all they can about these movements by vast numbers of birds. Both have plenty of material to feed their passion, for, far from following rigid rules, migrations are constantly evolving as routes and destinations change over time.

This book celebrates the wonder of migration, and the beauty of the migrants themselves. It also gives a summary of what is known of the reasons for the phenomenon.

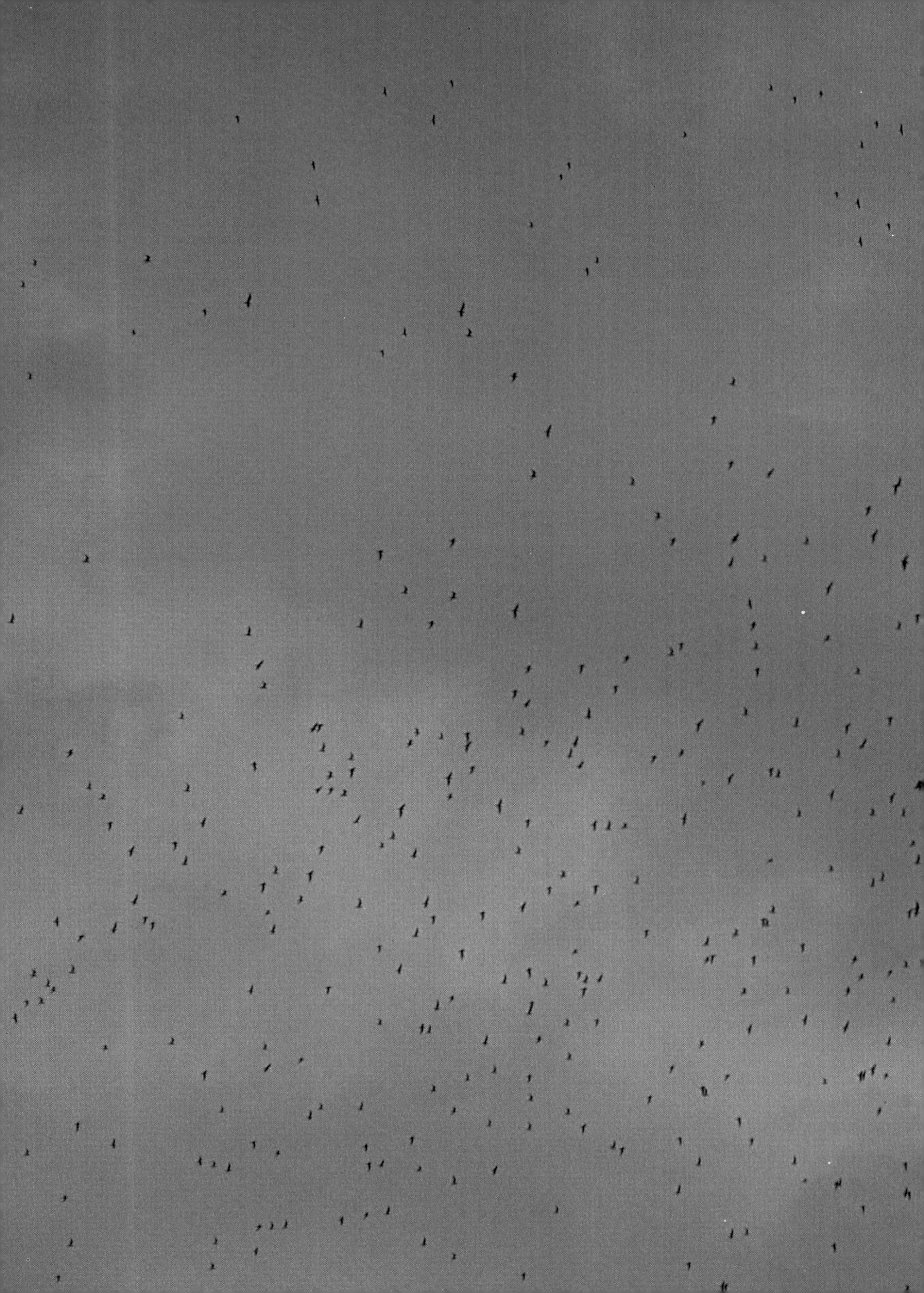

autumn

A sLight fading of the Leaves, an unaccustomed coolness in the early morning, a barely perceptible change in the light, and the smell of damp earth and mushrooms - for centuries these signs have announced the arrival of autumn to country people all over Europe.

The irresistibLe caLL of the south

When the first flocks of geese or cranes from the north began to streak across the sky, heading for milder regions, there was no longer any doubt: the balmy days had gone. This determined, vigorous movement bore - and still bears - witness to a will to live, rather than to disappear from view temporarily or permanently, as when plants shed their leaves, or when mammals and reptiles hibernate.

Now that we are less close to nature, few of us still think of raising our eyes to the sky in the hope of seeing what the philosopher Alain beautifully described as "the geometry in motion" of migrating birds. Fewer still among us are able to see or hear one of these groups of winged travellers.

And yet every autumn, from time immemorial, hundreds of millions of birds swarm across the continent. Still they pass high above us, overflying our cities and countryside, day and night. How does all this happen? And why do they set out?

Top left
Northern fulmar.

Right
Controlled power, calm, and utter beauty of form - whooper swans coming in to land.

Preceding pages
A flock flies past at a great height, a perfect example of the vast scale of autumn migration.

Five billion birds

Birds are not the only animals to travel great distances. Mammals and insects - from the African elephant to the American monarch butterfly - also undertake seasonal migrations. However, migration is more obvious among birds, because it can be seen in the sky and because it involves large numbers of individuals and species. Once they have finished nesting, most migratory birds travel south, over distances that vary from a few hundred to several thousand miles. This is known as "departure" or "postnuptial" migration because it takes place after breeding. Migratory birds leave their breeding areas essentially because weather conditions deteriorate and food supplies diminish or disappear.

"It was with inexpressible pleasure that I watched the return of the stormy season, the swans and woodpigeons passing overhead, the crows gathering in the meadow by the pool and going to roost at nightfall on the avenue's tallest oaks.**"**

François René de Chateaubriand

When a migrating bird crosses a given region, it is known as a "passage migrant". If it remains there in the winter, it is known as a "winter visitor". Once winter is over large numbers of these migrants head north once more towards their breeding grounds. This is known as the "return" or "prenuptial" migration.

However, the autumn movement of birds is far more spectacular than that in spring. The chief reason for this is simple: far more birds are on the move. It has been estimated that 5 billion European birds of about 200 species leave to overwinter in Africa. This substantial number represents only about a tenth of the world's total of migrating birds. The adults are joined by the vast contingent of young birds born since the spring. Moreover, the migrants' numbers are on the whole still not depleted. It will be a different story a few months later, when many will be missing on the return journey in the spring. Exhaustion, hunger, accidents, predators and, of course, shooting will have taken a very heavy toll.

There is a second reason why the autumnal migration is more noticeable. Although it is hard to generalise because of the many local variations, the busiest period lasts noticeably longer. The first movements can be observed at the end of July, whereas the last movements of any size take place towards the end of November. Between these times - from mid-August to mid-November - migrants are extremely active. The movements therefore take place over a good three months. In spring, on the other hand, most migration is concentrated over just two months: April and May.

Left
A great ribbon moving across the sky, this line of cormorants stretches and undulates, each bird seemingly chasing the one in front.

Right
Goldfinches fly south in a dancing, colourful band. Though less spectacular than the passage of flocks of geese or storks, the migration of small birds nevertheless involves hundreds of millions of individuals crossing Europe.

Following pages
Slicing the air with vigorous wingbeats, bean geese journey towards western Europe, forming a vast capital "V" in the sky. At the end of their journey the meadows and fields of stubble that welcome them every year await them.

Bizarre theories

The reasons why birds embark on journeys that cost them so much effort have always intrigued naturalists and others interested in natural phenomena. Despite this, the concept of migration was slow to gain acceptance. Indeed, until the beginning of the 19th century, scientists such as the great Cuvier believed the ancient explanation that the birds that disappeared with the approach of winter holed up in various crevices to hibernate. It was even thought, in defiance of the most elementary physiological principles, that swallows dug themselves into the mud of ponds to spend the winter there. However, certain less gullible or more observant minds - from Aristotle in the fourth century BC to Buffon in the 18th - speculated about, or even observed, the departure or wintering of certain species such as quails, storks and cranes.

Many theories have been advanced to explain the origins of bird migration. These must however remain conjectural, because we have no proof of the early movements of bird populations, even though we may have fossil evidence of the presence of a certain species in a given place and time. These discoveries lead us to suppose that the birds in question may have moved from one place to another with the seasons, but intellectual honesty demands that these remain no more than suppositions.

Driven by the ice ages

The fact remains that some of the reasons we suppose may lie at the root of migration - such as the need to adapt to local availability of food - can be applied to migratory birds on different continents. Others however are more closely linked to specifically European circumstances and to the continent's recent geological history.

Thus glacial periods - the latest of which began more than 70,000 years ago and ended some 15,000 years ago - played a large part in giving rise to the phenomenon of migration and in establishing its particular variations. Gradually driven further and further south over centuries by the advancing ice, many species discovered new, more southerly habitats to which they became adapted. Other, less mobile species were probably decimated by the increasingly harsh environment they faced. When the ice retreated, by contrast, birds were able to advance gradually northwards to nest on the "liberated" lands. However, they left these once nesting was over, chiefly because of the cold season that followed the warm.

> " oh that I had wings, wings, wings! as in Ruckert's song, to fly far with them to the golden sun and the green springtime! "

Théophile Gautier

Left
The great white egret's sinuous white shape can now be seen in places where it was once unknown - a sign of ongoing changes in its range.

Top right
Originating from Sweden or Finland, great flocks of migratory woodpigeons trail across the autumn sky. The great Iberian oak woods that guarantee plentiful food are still a long way off ...

Bottom right
Despite their great numbers, each white stork soaring in this thermal has its own space.

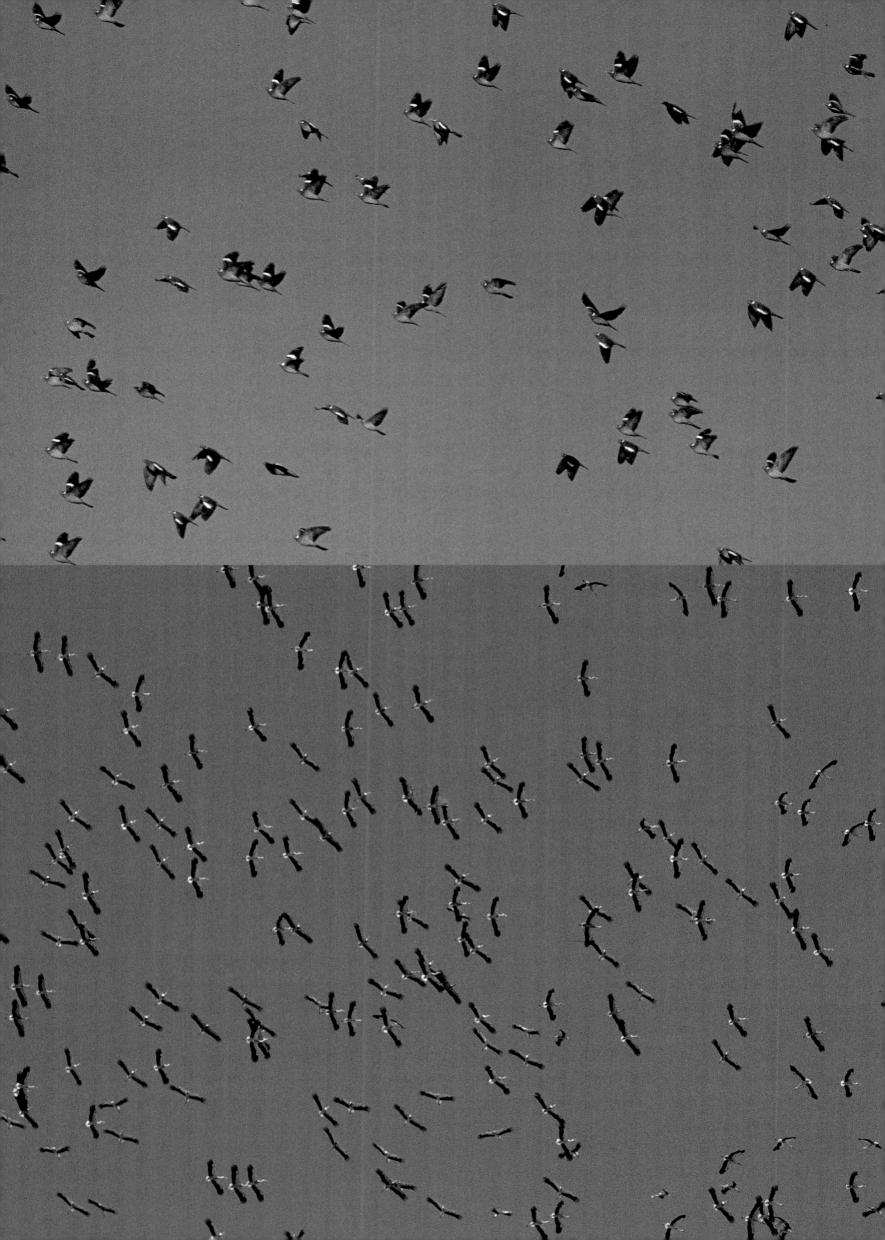

Most species therefore returned to more southerly regions, with which they had become familiar in the past. It is believed that some of the migratory movements we see today were set in train, or at least reinforced, by such conditions. It is none the less true that migration must have happened even before the alternation of glacial and interglacial periods, if only because of the distinct seasons that affected temperate and northern regions. Nevertheless, this is not the only factor likely to encourage migration, since this also happens among some of the birds that live in tropical regions.

The phenomenon of migration is rich in mysteries of all kinds, both as to its origins and mechanisms. Perhaps the biggest mystery of all is that of navigation, dealt with in the following chapter. It is easy to understand, for example, how certain species might be driven from a given region when climatic conditions make it inhospitable. It is hardly surprising that swallows, which are insectivorous, leave those regions which are short of suitable prey in winter. It might be argued that, this being so, what is surprising is that the birds feel the need to return north to nest when they might, perhaps, be able to do so in their winter quarters. But this would ignore the fact that competition for nest sites and food - indispensable to the survival of individuals and the perpetuation of the species - is fierce, and drives many birds constantly to seek living space.

An internal clock

If the case of the swallow is relatively clear, what are we to make of the cuckoo or the swift? These species generally leave their breeding regions when the temperature is still mild, and their food - insects - still plentiful. So, why leave?

It would be a mistake to assume that birds can foresee, or think about, imminent deterioration in their living conditions. In fact, they are subject to what is usually called their "inner clock". The fact that 20th century scientists have produced this explanation in no way diminishes the extraordinary nature of this biological mechanism. Without going into too much detail, it has been found that birds, like other living things, are equipped with a complex system which, via their endocrine glands, regulates their rhythm over the year. The moult (or moults) - that is, the replacement of a bird's feathers, which will be dealt with in the third part of this book - breeding, and, if applicable, migration, are all major phases in this annual cycle which follows a precise sequence. Part of this internal process is directly linked to day length.

To understand this phenomenon better, we need only think about how we feel ourselves as we emerge from winter and fine weather gradually takes over. This sensation of euphoria, this feeling that we are living again or being reborn, is largely the result of the lengthening days acting on our systems.

Above
The breeding season over, these shelducks gather for a time on the Waddenzee, on the coast of the Netherlands, to moult in peace. Once their wings have grown new feathers they can disperse to their wintering grounds.

Right
The holder of astonishing long-distance flying records, the Arctic tern is wonderfully streamlined. No doubt this helps to explain the bird's feats of endurance, in covering 19,000 miles every year.

Travelling companions

There are few solitary migrants among birds, most preferring to travel in groups. As one might expect, those that nest in colonies also migrate *en masse* or, at any rate, gather at resting places along the route. Terns, for example, like to travel in small or larger flocks. Bee-eaters also travel in groups, as do cormorants. However, many species that nest as single pairs or in a relatively dispersed way also gather together when the time comes to migrate. This applies to geese, ducks, cranes, storks, and the majority of waders such as sandpipers, redshanks and plovers, as well as certain raptors such as kites. These gatherings may be a result of the circumstances of migration itself: if there is only one convenient crossing point, it is natural for birds to gather there. But the birds may also instinctively seek the company of their own species for safety.

Hundreds of pairs of eyes are more likely to spot danger. Moreover, predators generally hesitate to attack large groups of birds, since they seem to find the large mass daunting and have difficulty singling out a victim.

This tendency to migrate in groups is a blessing in that it can produce astonishing, sometimes wonderful, sights: the slow passage of a "train" of white storks several hundred strong over the Straits of Gibraltar, the procession of dozens and dozens of honey buzzards - elegant raptors - effortlessly crossing a Pyrenean pass, or the passing throng of woodpigeons - the celebrated *palombes* of south-west France, which form what are locally known as "blue swarms" - en route for Spain with its plentiful food. Other such sights include thousands of cranes crossing northern Germany or the lakes in the Champagne region of north-east France. The list is endless.

Above left
Despite its small size - barely bigger than a sparrow - the Temminck's stint is an accomplished traveller, able to reach as far as equatorial Africa after nesting in the Arctic tundra.

Above right
The wood sandpiper nests slightly further south than the Temminck's stint, but its migration may take it as far as southern Africa.

Right
They may appear disorderly but these geese, which have just taken flight, will soon gather into regular formations, as they have for millennia. Furthermore, in the apparent confusion family groups stick close together.

once the phenomenon of migration had been identified
and accepted as fact, a system was needed to find out with reasonable
accuracy where migratory birds went, and by what routes. The obvious
answer was to mark individual birds so that they could be identified.

And still they Leave ...

Several methods were tried and, after much trial and error, the best way
of "marking" a bird was found to be by ringing. The pioneer of ringing
was Christian Mortensen, a Danish schoolteacher, who in the last years
of the 19th century had the idea of using aluminium rings bearing
engraved data.

Top left
Gaudy, with its parrot-like
brightly coloured bill during
the breeding season, the
puffin leaves the cliffs after
nesting and heads for the
open sea. It then spends the
winter in the North Atlantic,
buffeted by the waves for
several months on end.

Right
This common crane seems to
be giving out a clarion call
before setting off on its
voyage. Crane migratory
routes have been known in
detail for a long time: their
size hardly lets them pass
unnoticed, and they have
long been prized by hunters
who became well acquainted
with their habits and
especially their routes.

rings on their legs

The idea is simple: each migratory bird - usually caught with a light net, carefully sited to ensnare a specimen of the species concerned - is fitted with a leg-ring bearing a code. When the bird is recaptured - known as a "controlling" - or found dead - a "recovery" - its departure and arrival points can be established, but of course the details of its route remain unknown. This is especially true when there is a long time between the dates of the bird's initial capture and of its controlling or recovery, during which time it may have wandered considerably.

For some decades now new methods have been used to plot bird movements without the need to recapture the birds. These systems make extensive use of coloured rings made from synthetic materials such as Kevlar or PVC, attached to one or both legs, and whose combination of colours allows identification. Larger species are fitted with rings with a surface big enough to be marked with an alphanumeric code that can be read at a distance. Other devices include coloured plastic tags fixed to the wing, and bands or "ties" fixed to the neck. These ultra-light gadgets are not believed to trouble the birds, such as herons or geese, that are fitted with them. The great advantage of a number readable from a distance is that it

allows the precise identity, age, and origin of a given bird to be established without capture. Apart from the scientific interest of recording marked birds, it is a great pleasure for a birdwatcher to learn for example - after consulting the ringing team, a matter now made easier by the Internet - that that brent goose he has just spotted is at least 11 years old and has covered the impressive distance from the Taimyr Peninsula in central Siberia to a small bay in the Côtes-d'Armor in northern Brittany.

In just a few decades - despite the low rate of controlling or recovery of ringed birds of many species (in general 1 to 4% for small passerines, but sometimes as low as 0.5%) - the many ringing programmes undertaken around the world have gathered a vast amount of information about bird migratory movements, which are now much better known and understood. Indeed, many species are no longer studied through ringing programmes because researchers feel they have gleaned enough data and see no justification for needlessly exposing birds to the small, but real, risks of being ringed.

> " ... the multitude of unchanging shapes orders itself through movement, each individual gliding into its neighbour's slipstream and happily finding its shape traced there. As for the intricacies of this flying technique, we badly need an essay on the subject by a goose qualified in geometry; but these powerful navigators don't stop to think about it. "
>
> **Alain**

Left
The large rings attached to birds' legs may diminish their natural beauty, but they are nonetheless invaluable to our understanding of the movements of migratory species.

Right
In flight, birds give no indication of where they have come from. Only once they have alighted might they reveal their secret, thanks to their rings bearing information that can be read at a distance.

Recent advances in transmitter technology, especially their miniaturisation, have greatly increased our knowledge of migratory movements. For example, tiny Argos beacons attached to storks or albatrosses have allowed the long-distance movements of these large birds to be plotted in detail.

An orienteering race

As migrations came to be better understood, thanks initially to ringing, specialists still could not understand how birds navigated on their journeys. Although much more is known about this today – and more is coming to light all the time – many mysteries, or at least questions, remain.

What we do know is that birds use different means of navigating, probably combining them in a variety of ways: these include using the stars and the sun, and memorising topography. They are also believed to use less "Cartesian" methods such as sensing the earth's magnetic field, judging the direction of light – and thus the sun's position – even when the sky is overcast, and perceiving infrasound waves, which, for example, would allow them to navigate using the noise of waves, even from a distant sea. Some researchers have gone further, speculating that birds may have a particularly keen sense of smell. This theory, however, could only apply to the few families of birds where this sense is sufficiently highly developed, such as the "tubenose" order of seabirds (albatrosses, shearwaters, petrels and relatives).

However, these explanations cannot account for everything. A particularly astonishing example – among many others – is that of young cuckoos, which travel unprompted from Europe, where they were hatched, to tropical Africa, where they will spend the winter, even though their parents have already left, often several weeks earlier. It is hard to talk about learning or memory in such a case. How do these young cuckoos know their destination, and the route?

No doubt our advancing knowledge will one day allow us to solve these mysteries. For the moment, however, we can only wonder, and marvel, at such feats.

Above
From an altitude of several hundred feet a landscape looks very different from how it does at ground level. In good weather, migratory birds have a sort of map in which landmarks such as coastlines and wetlands act as guides.

Left
The cuckoo loudly proclaims its arrival in the spring, making liberal use of a call that has earned it an evocative name in every country it visits. By contrast, when it leaves it does so silently, fading furtively away with the passing of summer.

Right
Vast numbers of pink-footed geese from Spitzbergen, Iceland, or Greenland assemble for their autumn voyage, flying over the sea for hours on end.

ALThough some species - such as penquins, whose flipper-like wings can only be used for swimming - can migrate without flying, the vast majority of birds take to the air to do so. It is, indeed, a highly practical mode of travel.

FuLLy equipped for fLight

Flight - at least, flapping flight, for soaring is another matter - is an energy-expensive mode of locomotion; but this disadvantage is more than compensated for by the speed of travel, which means that the time devoted to this intense effort is relatively short.

Top left
The black stork is an accomplished soarer, whose 6½-feet wingspan carries it effortlessly to Africa.

Right
Small passerines, with a wingspan often of less than 12 inches, can still travel long distances at their own pace. Every autumn this yellow wagtail (northern race) flies from western Siberia to tropical Africa.

Lightness and power

It is hardly surprising that birds possess all the characteristics necessary for flight; nevertheless, we cannot but admire these properties. In essence, birds have the advantages of lightness, power, and a number of physiological adaptations specific to flying.

Their lightness is above all due to a skeleton whose bones not only lack bulk but are, most importantly, pneumatic: as well as being hollow and strengthened by a lattice of "cross-braces", they contain "air sacs". These pockets, linked to the bird's respiratory system, are distributed throughout the body within certain bones. Birds also lack a bladder, their highly concentrated urine taking the form of a whitish paste. Finally, their skin, devoid of sweat glands, is extremely thin.

Feathers also offer a remarkable compromise between lightness, strength, and efficiency. Firmly rooted in the wing, they make flight possible and, as watertight as they are airtight, ensure effective insulation.

The main flight muscles are the pectorals, which alone account for about a third of a bird's weight. Their high level of development is linked to that of the keel, the blade-like protrusion that juts forward from the sternum and which is reduced or absent in non-flying birds.

Finally the lungs, linked by the pneumatic sacs, are protected by an incompressible ribcage, which prevents them from being crushed when the wings beat downwards in flight.

Above, top
Flying geese exude a feeling of power. Endowed with great stamina, they have long, pointed wings powered by bulky pectoral muscles.

Above, bottom
Common cranes ascend effortlessly to heights where they become invisible to the naked eye.

Right
Waders fly fast, with shallow, rapid wingbeats. Once under way these spotted redshanks can cruise at some 37 miles without excessive effort.

cruising altitude

Migratory birds travel at different heights, but most cruise at a few hundred feet. They generally fly higher at night, and when flying north in the spring. However, it is estimated that about a third of birds travel at between 3,300 and 6,600 feet, while a much smaller proportion goes higher still. Often, the majority fly at about 1,300 feet by day and at about 2,300 to 3,000 feet by night. These figures may vary depending on relief, the presence or absence of stretches of water, wind, type of flight and, of course, on the species.

Wind is an important factor in migration. If it is a gentle, following wind, birds take advantage of it and fly higher. In a head wind, on the other hand, they drop their altitude. If the wind is too strong, many birds, especially the smaller species, fly so low that they almost skim the ground. Finally, in very strong winds, especially if from an unfavourable direction, many of the migrants less well equipped to deal with severe conditions break off their journey rather than risk being blown off course, especially out to sea.

As mentioned, different species migrate at different altitudes. Passerines, which are mostly small, fly relatively low. Thus it is not unusual to see certain finches or larks crossing cities at roof height. By contrast, ducks, geese, raptors and waders fly higher.

Flying at height has a number of significant advantages. The greater the altitude, the cooler the air, and the easier it is to progress because friction is lower. Wind, too, tends to be less changeable higher up, while a further benefit is that vertical turbulence is usually less. Finally, the risk of meeting a predator is far lower.

Birds that use flapping flight tend to fly lower than species that soar, which are a special case. These birds, notably storks and raptors, take advantage of rising thermal currents to soar effortlessly up to the right height for their journey. Once they have reached the top of their climb, these master gliders, endowed with a large wing surface area, begin to descend at an angle in the direction of their journey, until they meet another thermal which will bear them aloft again, sometimes to several thousand feet above sea level.

The records for altitude are held by geese and vultures, which have been recorded at 30,000 feet and 36,000 feet respectively: the former above the Himalayas, the latter in Africa. We might well wonder how warm-blooded creatures can stand such heights, where temperatures are very low, and oxygen scarce. The explanation, which has only been discovered recently, is that the blood of these birds contains different types of haemoglobin, with varying densities and oxygen-bearing properties. It is as if a car engine contained oil whose lubricating properties varied according to the conditions.

FLight speed and endurance

It is difficult to measure the flight speed of a given bird. However, whatever the methods used, the figures recorded are lower than perhaps might be expected. In general, migrating birds fly at a few tens of miles per hour: 19-25 miles per hour for small birds, and up to 37-50 miles per hour for larger ones. In very unusual conditions, when they are aided by fast air currents, some birds such as swans or geese can reach much higher speeds, sometimes more than 93 miles per hour.

Contrary to what we might think, most migratory birds do not fly for long periods - quite the opposite. In many cases, especially among smaller passerines, only two, three, or four hours per day or night are spent travelling.

Some species are capable of flying for long periods when forced, as when they have to cross a wide stretch of water or a desert. Certain waders, such as knots, nest in the Arctic and migrate in large numbers across the North Atlantic to reach Iceland or the British Isles. This involves flying for at least 100 hours - an astonishing feat. The official endurance record was set recently by a Bewick's swan, which made a prodigious single flight of 1,100 miles, from Russia to Lithuania.

Above, from top
Hobby, glaucous gull, and white-tailed eagle. Travellers great and small, migratory birds have an ability that has evolved slowly over millions of years.

Right
Montagu's harrier, with its golden eyes (above) or little egret (below) can travel considerable distances without flapping their wings, thus saving large amounts of energy.

" To flee! to flee to that far place! I feel the birds are drunk among the foreign sea-foam and the skies! "

Stéphane Mallarmé

IT seems incredible to many that migratory birds travel by night. People can understand that nocturnal species - owls first and foremost, but also nightjars - can fly in complete darkness without difficulty.

Journey to the end of the night

However, it can seem inconceivable that diurnal birds manage to cross the skies at dead of night. Yet they do, emphatically so: in fact more migrate by night than by day, both in terms of number of species and in total numbers of individuals. Why should this be?

Top left
Geese and ducks, such as this greylag goose, are nocturnal migrants that can also travel by day.

Right
Answering the call of the night, these common cranes make an impressive silhouette against the evening sky.

Dancing in the dark

In addition to exclusively or mainly nocturnal migrants such as insectivorous passerines (warblers, flycatchers, thrushes, robins), waders, ducks, and even the quail and the cuckoo, some mostly diurnal migrants such as pipits and larks are equally at home travelling by night. By contrast, finches - seed-eating passerines that include chaffinches, bramblings and linnets - and swallows migrate almost exclusively by day. Similarly, large soaring birds such as storks, pelicans and raptors need thermals to fly long distances while saving their energy. Thus, in autumn and in spring, while we sleep, a few hundred feet above our heads the sky is traversed by millions of birds, forming an amazing aerial ballet which, despite its vast size, passes mostly unnoticed.

In the safety of the night

Paradoxically, night flying offers many advantages. For a start, it saves a lot of time. In theory, at least, a diurnal migrant has less time to travel: it must use part of the day for feeding, at the expense of travelling time, since the night is devoted to resting. By contrast, a nocturnal migrant uses part of the night for flying, then alights to rest and look for food during the day.

Moreover, flying conditions are better at night, which makes more efficient use of a bird's muscles and thus saves energy. Night air is cooler, and easier to penetrate. In general, winds are lighter and less changeable. Finally, vertical turbulence is rarer and less intense. There are physiological advantages too: at night the birds are less at risk from overheating and dehydration, dangers linked to sustained muscular activity.

Another reassuring aspect of night flight is the almost complete absence of predators, especially if flying sufficiently high. To make their journey safer still, night migrants fly at heights that originally avoided natural obstacles and now avoid - though not always - death traps such as electric cables.

Above
Water birds are often active at night, whether feeding or travelling. While on migration however they may also rest for a while, like these common cranes. Wading in water, they feel safe from attack by land-based predators.

Right
Slipping past against a background of clouds rimmed with the gold of the setting sun, these common cranes are perhaps settling down for a night of flying, during which they will cover several hundred miles.

Following pages
With the night sky to themselves, common cranes fly over a large area of wetland, which may provide some of them with an ideal stopover point.

"Birds, the most ardently alive of our fellow-creatures, follow their singular destiny to the end of the dying day. Migrants, possessed by the swelling sun, they travel by night because the days are too short to encompass their activity. When the moon is as grey as the mistletoe of Gaul, their ghosts inhabit the prophecies of the night."

Saint-John Perse

the tracks of the stars

As we have seen, birds have a stellar compass, among other navigational aids. Many experiments have proved the existence of this precious asset. It is believed that this knowledge of the heavens is developed, at least in part, by young birds while still in the nest, when they acquire the habit of observing both the apparent movement of the stars and the motionlessness of the Pole Star. Once firmly memorised, these observations are precious pointers when the time comes to embark on their first migration. To these data are added elements that are harder to define but which for convenience we can call instinct.

However, this stellar compass is only one of several systems that allow birds to find their way at dead of night: there are believed to be others that can be used in combination with it. Thus, when the stars are hidden by cloud, birds can navigate according to the point where the sun disappeared over the horizon, or by finding their position using natural sounds such as that of waves, indicating in which direction the sea lies, or of the wind rushing into a valley.

cries in the night

Night migrants are very talkative. Diurnal ones are too, but calling is thought to be less important, since they can see as well as hear each other. Night migrants regularly utter specific calls, often quite different from those they make by day, chiefly, it is thought, to maintain contact with their companions, and even with other species. Of course, such safety is relative, for the calls may come from birds that have taken a wrong course, for example by heading towards the sea. Possibly such calls can be modulated to warn of danger.

emergency landing

Birds certainly prefer to set off on clear nights but, if the weather worsens a few hours after they have taken off, especially if there is fog or rain, they are forced to take emergency measures. This consists in alighting as soon as possible, often at random. Thus migrants can sometimes be seen early in the morning in urban green spaces, or even on the roofs of apartment blocks, if they have been caught by bad weather while flying over a built-up area. The dangers they face are then considerably increased. These forced landings sometimes involve such large numbers of birds that they are known as "falls". There have been cases where people have woken up to find the ground thickly strewn with birds in various states of exhaustion.

Above
Most small passerines migrate by night. An avowed lover of darkness, the red-breasted flycatcher travels from eastern Europe to India in a series of night flights. Its large eyes are a sign of its ability to make the most of feeble light.

Right
These common cranes appear to be following mysterious heavenly signs.

Tens upon tens of millions of birds set off on migration. It is not surprising, therefore, that these travellers pass through almost everywhere, while avoiding the least hospitable regions.

skyways

In principle, it might appear that a bird can fly anywhere, over any type of terrain. However, the truth is rather more complex. First of all, land birds such as warblers, thrushes, turtle doves and quails, avoid flying over open water, especially large stretches, whenever possible. Conversely, birds that live on water are often very reluctant to fly over large land masses. However, this aversion is less marked than the opposite tendency in land birds. Herons, waders, and ducks, for example, fly over land - even over vast stretches of desert - without hesitation. Those most averse to crossing large continental land masses are seabirds, some of which, such as petrels and shearwaters, highly dependent on their habitat, only visit land if they are blown there by violent storms, usually westerlies. Some seabirds however, such as skuas, overfly land at a great height, especially when heading to their northerly breeding grounds in the spring.

Top left
Like most large soaring birds, the black kite is reluctant to overfly large stretches of water, which are a poor source of the thermals it needs to wheel effortlessly on its outstretched wings.

Right
While migrating, common cranes are creatures of very definite habits. Their almost unchanging routes lead them from one traditional staging post to the next.

clear preferences

Deserts and high mountain massifs are two types of terrain that migrants avoid overflying wherever possible. Heavily wooded areas, or open ones devoid of all but the lowest vegetation, may attract or put off migrants, depending on the species. A wader such as a curlew, for example, feels safe passing over large expanses of agricultural land, but will tend to avoid flying over a large forest. By contrast, a jay will remain under cover as much as possible and only reluctantly launch into crossing an area devoid of vegetation where, as a relatively poor flyer, it is temporarily very vulnerable.

The combination of these preferences and aversions means that certain places, such as valleys, spits of land protruding out to sea, and mountain passes, are favoured by migrants.

All to the front

Ornithologists distinguish between three different types of migration, in terms of the breadth of the front across which birds advance.

Many birds, especially passerines (which account for the majority of migrants) but also waders and ducks, advance across a very broad front, although this excludes, as indicated above, mountainous areas, deserts, and wide stretches of sea. With certain exceptions, these migrants travel without showing clear preference to any particular type of terrain.

Other species take a much more precise route. The best example of this is the common crane. These large birds (to which a later chapter is devoted) cross France diagonally, across a front which in theory barely exceeds 190 miles in width and, in practice, is limited to a few tens of miles. Often, however, some of these birds depart from their usual route and can be observed in regions where they are never normally seen.

The last type of migration warrants detailed examination, if only because it produces spectacular concentrations of birds.

Left
Woodland passerines, such as this chaffinch, find forest regions welcoming and have no fears when overflying them on migration – unlike species foreign to this type of habitat.

Right
Most wooded areas suit the pied flycatcher (here a male in summer plumage), which makes free use of urban green spaces on its journeys across Europe.

Gathering at the straits

Large soaring birds have already been mentioned in the chapter dealing with flying techniques. Certain species - storks, pelicans, raptors - rely on being carried on warm air currents known as "thermals", that allow them to gain height. Since these rising air masses are produced only above land warmed by the sun, the migrants that use them cannot fly over stretches of sea unless these are narrow. In order to cross from Europe to Africa or vice versa, these migrants therefore gather at a few places where the sea is at its narrowest. The two most famous are the Straits of Gibraltar, between Spain and Morocco, and the Bosporus, between European and Asian Turkey. Another well-known crossing point is the strait separating Denmark from southern Sweden, especially around the Falsterbo area on the Swedish side. Passerines and raptors, especially from much of Scandinavia, gather there in autumn on their way, initially, to central Europe.

While at Gibraltar and the Bosporus birds must cross the sea - albeit narrow at these points - at Eilat they take advantage of the narrow isthmus that separates the Red Sea from the Mediterranean. There, in southern Israel, vast multitudes of birds, including astonishing numbers of raptors, pass from Europe and western Asia to Africa, and vice versa. Caught between the Sinai, the Arabian Desert, and the Red Sea, they have no other route - hence these spectacular concentrations.

Aside from these important sites, there are others worthy of interest even if they do not play such a fundamental role. Thus certain Mediterranean islands are useful staging posts for many birds which, while not so dependent on overflying land as storks or large raptors, are nevertheless grateful for the chance to make a stopover between Europe and Africa. Malta and Cyprus, for example, play host to large numbers of migrants.

Crossing points that are bottlenecks for land birds are also convenient "gates" for certain seabirds such as shearwaters and terns. Furthermore, some other places are little used by land birds but are heavily frequented by water birds, especially seabirds such as divers, auks, skuas and sea ducks (scoters, eiders). A good example of such a passage point is the narrows between the Channel and the North Sea, a route chosen by many birds which prefer it to flying over land.

> " It draws near,
> it is coming:
> the goose tribe.
> Like arrows all,
> their necks outstretched,
> hurtling ever faster in their
> wild flight, they pass,
> whipping the air with their
> whistling wings."
>
> **Guy de Maupassant**

Left
When they come to a large stretch of water such as a lake or an arm of the sea, some migrants, especially small passerines, prefer to take the long way round it rather than strike out over the inhospitable waters.

Right
Even though their rapid, high-altitude flight means that they pass over most types of habitat, greylag geese, such as this particularly quarrelsome group, ensure that wetland areas occur at regular intervals along their migration routes. Ducks and waders adopt the same strategy, dictated by their ecology.

migratory birds rarely make very long journeys without stopping. For obvious physiological reasons, they generally break the journeys up into stages.

ports of call

Without these stops, migration itself would be hard to detect. For example, how would we know about the nocturnal migration of pintails, without having seen a group of these ducks resting by day on a pool?

Top left
A rare migrant: the glossy ibis.

Right
Although waders spend much of their time feeding, they also need to spend some time resting, in order to save energy and be up to continuing their journey. This dunlin has taken a brief break from combing the salt flats: its beak buried in its feathers allows it to conserve as much heat as possible, while its half-open eye shows that the little wader remains on the alert.

enforced stops

Like many other animals, migrating birds use stored fat as an energy store to "fuel" them along the way. This fuel is mostly used up by the intense muscular activity that migration demands. The principle is a mathematical one: a certain amount of fat "burned" equals a certain amount of muscular energy produced (an aspect which will be dealt with in detail in the last chapter). It follows that once the fat is all, or almost all, used up, the body can produce no more muscular power. Once it has reached this stage, a migrating bird must stop to replenish its reserves by feeding. If - at least in the case of land birds - it can alight in a favourable spot it has a good chance of doing this. If, on the other hand, the environment is unfavourable or even hostile - such as a stretch

of water - the situation becomes difficult, even serious. As a rule, birds ensure that they do not get into a position where they face disaster just to replenish their energy reserves. But this depends on the weather not turning severe. If they encounter strong winds, persistent rain, or snow in the higher latitudes, the problem takes on different dimensions. In other words, a given quantity of energy will not be enough to produce the same results as under normal conditions, since part of it is needed to combat wind, cold, or both. Because of these unforeseen factors, theoretical calculations are never precisely confirmed in nature.

travellers with different timetables

Many migrants thus devote only a part of the day or night to travelling, at least while migration is at its height. Sometimes they only travel a few tens of miles in a day. Moreover, when a site is especially safe or well stocked with food, they will extend their stay there, resting for several days. At this relatively leisurely pace migration can take weeks or even months, especially in autumn. In the most extreme cases migration can be seen as a continuous process: having arrived at the southern end of their journey, some species waste little time before setting off back to the regions where they will nest the following spring.

Left
Chance events on migration sometimes lead birds to stop - occasionally very briefly - in unusual, unsuitable places. This blackcap, which has wandered away from cover, would normally stay among trees and bushes, where it can feed on insects from foliage.

Right
The common sandpiper is not choosy as to where it stops off on migration as long as the site has water. It combs the banks of rivers and canals, the shores of pools and even artificial lakes, sometimes in town centres.

Many birds that winter in southern Africa leave Europe in August or September, but do not reach their destination until November or December. In general the autumn journey takes about two and a half to three months for large migrants travelling the longest distances, two-thirds of that for medium-range migrants, and a month for those making the shortest journeys.

Of course, alongside these migrants, which might be described as relaxed, there are others that can cover the miles at a truly astonishing rate. There are plenty of examples: there have been cases of birds covering some 620 miles in little more than 24 hours. The prowess of birds from Canada or Greenland crossing vast stretches of ocean, with no prospect of alighting before reaching Europe, has already been mentioned.

keeping faith with a place

Information gathered through ringing has gradually brought to light a phenomenon which had been unsuspected or, at any rate, whose range and implications had never been appreciated. Birds are often highly faithful, not only to the places where they spend the winter but to the routes they take to get there. Thus migratory ducks such as the garganey may stop at the same pools year after year, before wintering in the same

African marshes. Similarly, marsh-dwelling warblers often frequent the same European reedbeds along their southward migration, and return to the same habitats once they have reached Africa. Waders and raptors display similar devotion, which suggests remarkable powers of memory.

There are plenty of examples of such astonishing powers. A cormorant from Denmark was seen repeatedly at the same place in northern France, not just on the same tree, but on the same branch. Naturally, species that tend to live longer are more likely to develop such habits, but they are not the only ones to do so. And although ringing is the favoured way of checking such behaviour, it is not the sole method: some observers can identify a bird by some distinguishing characteristic, and thus recognise it over the years.

Being faithful to stopover or wintering sites carries certain risks for birds. They must be able to adapt to any changes in these habitats, or see their chances of survival reduced. This subject will be dealt with later in the book.

Left
Many birds that are dependent on coastal areas stay close to them while migrating: oystercatchers are, above all, coastal waders that like to feed and rest on salt flats exposed at low tide.

Top right
Migrants must take great care of their feathers. These ringed plovers devote considerable time each day to preening.

Bottom right
While the sanderling in the centre of the picture is stretching its left wing and leg in a typical gesture of relaxation, the one on the right has suddenly succumbed to fatigue and is dozing, motionless.

Following pages
Flocks of migrants often form beautiful patterns. At high tide, waders must find resting places where they can calmly wait until the next low tide exposes the mud flats where they can feed. Small waders like to congregate with their own species, but among these dunlin (one of which has a coloured ring) an "intruder" in the shape of a ringed plover can be seen at the bottom centre of the picture.

The right Lodgings

In its broadest sense the term "migration stop" can be applied to any welcoming spot, where one or more migrant birds stay for a few hours or a few days. Certain environments may lend themselves more than others to playing host to migrants but, for example, a temporary pool among fields where curlews or redshanks linger for an hour or two certainly counts as a migration stop. Indeed, when weather deteriorates, migrants may be forced to alight almost anywhere. A leaf warbler caught in a violent downpour at night might be forced to take refuge in an overgrown ditch or a city park, while under normal conditions it would choose a wooded area containing deciduous trees and plenty of undergrowth. Nevertheless, the ditch or park would constitute a substitute, but vital, migration stop.

When they can, however, migrants seek out the habitats that suit them best. In this respect the less specialised or demanding species are better served. A blackcap, song thrush or robin has at its disposal a wide range of suitable wooded sites. The same is not true of species with more specialised habitat needs, such as birds of wetlands, whether these be ponds or marshes. Rails, crakes, many ducks, snipe and bitterns, for example, depend heavily on finding suitable wet areas.

Seabirds however have no problems finding stopover places on migration because, although they require a special habitat - especially pelagic species, which keep to the open sea - when necessary they can alight on the water. Their only imperative is to find areas with enough fish to keep them fed on their journey.

spectacular gatherings

All over Europe, the most attractive sites draw concentrations of birds, sometimes in vast numbers. These are for the most part wetland areas, often close to the sea. The most famous of these are of crucial international importance for bird conservation. They include the Guadalquivir marshes and Ebro delta in Spain, the Neusiedlersee in Austria, the Camargue and the Golfe du Morbihan in France, and the Danube delta in Romania. As well as these places - which are special not just for birds but for birdwatchers and ornithologists - there are many less spectacular but ecologically very important sites that are worthy of study. Like those above, these are often wet areas: pools, marshes, coastal lagoons and plains liable to flooding. Estuaries, sheltered coasts that have mudflats at low tide and offer resting places such as islands, islets, or rocks at high tide are also very attractive to birds during migration, and may remain so over the winter. In addition to wetlands, migrating birds are also strongly attracted to areas that offer a mixture of habitats.

Above, top
Migratory water birds depend to varying degrees on wetland sites.

Above, bottom
This is where they can find the often highly specific food they need to regain their strength. Here, too, they feel safest and so can rest in comfort. Stretches of water are not the only wet areas sought out by migrants, which also appreciate flooded meadows or wet grassland.

Right
En route to Africa, these white storks have gathered in a field of stubble in southern Andalucia. After this stop, which they will use to catch plenty of large insects, they will have to tackle the crossing of the Straits of Gibraltar, to reach the Moroccan shore.

" musical storks, lovers of church bells, oh, how sad it is that you cannot sing! ... I love you sweetly, for I see you at rest, with the soul of Egypt in your hearts. "

Federico García Lorca

Thus, especially for passerines, a mixture of bushy and open areas is attractive, as is agricultural land alternating with small copses. By contrast, birds such as geese, lapwings, golden plovers, or cranes need above all wide open spaces where they can see far, thus offering their flocks maximum security.

The gatherings of birds observed at the best-known sites are often astonishing, sometimes even awe-inspiring. To watch thousands or even tens of thousands of waders - sandpipers, godwits, redshanks - swirling through the air in perfect synchrony, to see coastal rocks literally buried by a multitude of plovers, turnstones, or oystercatchers, or to come upon a pool totally covered with ducks: these are unforgettable sights for those lucky enough to witness them. The knowledge that these birds have come from distant lands, that they have often travelled through the night, and that soon they will be elsewhere, usually in a landscape totally different from the one where they were born, adds to the stunning scale and beauty of the spectacle.

Degradation of suitable habitats

Changes to much of Europe's landscape, which reached unprecedented rates during the 20th century, have affected

many migratory birds and even endangered some. On a local level this has affected small areas of particular types of habitat; more generally it has severely reduced the number and size of wetlands. In the first case the gradual disappearance of hedges, small lowland copses, and nourishing wild grassland has affected above all migrating passerines looking for a place to stop and regain their strength. In the second, widespread land drainage has reduced the wetlands essential to a large number of species - herons, waders, ducks, rails and crakes, and marsh-dwelling passerines (warblers) - to the bare minimum they need to survive.

Sadly, other factors have made these problems worse. The use of pesticides has wiped out many insects - especially the larger insects on which certain bird species depend - over vast areas, with serious consequences for birds such as hoopoes, scops owls and shrikes, whose diet consists largely of such insects. The replacement of mixed deciduous woodland by thick, sterile blankets of conifers, or the planting of the latter in areas which were previously ecologically far more diverse, have deprived many species of places where they could stop under the most favourable conditions while migrating.

> " In the evening I would row out alone over the lake, guiding my boat among the reeds and the great floating leaves of water lilies. There the swallows gathered, preparing to leave our latitudes. "
>
> **François René de Chateaubriand**

Left
The magnificent silhouette of the Montagu's harrier (here, a male) is no mere pointless elegance. Its streamlined shape makes it an efficient hunter that catches small rodents in open country - the type of terrain where harriers like to stop on migration to Africa.

Right
A fluttering flock of black-headed gulls brings the lake to life in a great flurry of wings. Such large numbers of these birds are usually seen only at the most attractive sites.

GLimmers of hope

Thankfully, all is not lost in the battle to preserve stopover sites for migratory birds. Alongside the existing network of established reserves, efforts have been made here and there in Europe to restore places where migrant birds can shelter under the best possible conditions, or even to create new sites for this purpose. In south-west France, the extensive Marais d'Orx, formerly drained and turned into maize fields, has once again found its true vocation as a Mecca for birds travelling along the Atlantic seaboard. In Britain, many wetlands have been created, notably on former industrial sites or harbours, sometimes in the middle of urban areas. The Netherlands, too, contains reserves which, though small, are very numerous and offer attractive stopover points.

Although private individuals can rarely create wetlands of any size, anyone with a garden can help migrants in a small way. The birds that benefit will above all be passerines such as warblers, robins, and flycatchers. Keeping a corner of the garden in a more natural and wilder state than the rest favours the presence of small insects, which passerines eat during their stopover. Planting or promoting the growth of berry-bearing shrubs, which offer plentiful food to many species, is another excellent step. These measures can seem trivial;

however, enabling a warbler to regain its strength and continue its journey is to improve – perhaps decisively – its chances of breeding next spring, and therefore contributes to the conservation of that species.

Left
While natural pools (far left) fit in better with the landscape and are a great biodiversity resource, artificial lakes (left) may sometimes become integrated with their surroundings and achieve ecological balance. They can then become attractive sites for wetland-loving migrants, as long as parts of them are set aside for such birds.

Right
The range of species and the numbers of birds are both indicators of the quality of stopover points. The best – often nature reserves – are those where the greatest safety is combined with plentiful food. Here greylag geese rub shoulders with a snow goose, wigeons, teals, mallards and coots, the last distinguished by their sooty plumage.

Following pages
Its beak buried in its plumage, this pochard can rest for a few moments in the security of a safe stopover point.

Fortunately, at the end of the breeding season a large proportion of migratory birds reach their intended destination. On the dangerous journey, hardened adults fare better than inexperienced youngsters.

vagrants

Among the many dangers that migrants face, going off course is naturally one of the most serious. This can be caused by a bird's navigation system becoming confused, or by external factors such as strong unfavourable winds. Although birds can compensate up to a point for the sideways drift caused by a strong wind, once this becomes too strong there is nothing they can do. Every autumn, whenever strong easterly winds dominate, some birds – usually small passerines – are blown out to sea while crossing the Straits of Gibraltar. Unusual birds also turn up in Europe during the migration season. Rare, non-breeding species, known as "vagrants", attract huge interest from birdwatchers, who flock to see a particularly rare migrant.

Top left
The little auk (seen here in breeding plumage) nests in its millions on the Arctic coast, and is a rare visitor to western Europe in autumn or winter. It stays out to sea, but strong westerly gales have been known to blow this small bird, which weighs just 3½ ounces or so, inland.

Right
Photographed on the Île de Sein, off Brittany, this paddyfield warbler from central Asia is far from its usual winter haunts in India.

A season for surprises

Vagrants to Europe during the autumn migration come either from Asia (usually northern Asia – that is, Siberia), or from North America. The former are brought by strong easterly winds, the latter by deep depressions moving from west to east. Careful scrutiny of meteorological maps allows the arrival of such vagrants to be forecast, and when conditions are right, birdwatchers rush to places known to attract them. The Scilly Isles in Britain and the island of Ouessant in France are strategically positioned on the edge of Europe: they are often the first land encountered by North American migrants after crossing the Atlantic – and the last land where migrants from Asia can alight. Birdwatchers there might chance upon a black or yellow-billed cuckoo (North American species) or spot an Isabelline shrike that has wandered far from its native central Asia.

In addition to these special places, there are others where migrants tend to congregate, such as spits of land by the sea, and straits, but vagrants can in theory turn up anywhere, which gives an added spice to birdwatching. This is all the more so because, as well as true vagrants, which are foreign to Europe, there are others that are unusual in a given country. Thus, while it is common enough to see a rough-legged buzzard in Norway, this raptor becomes a rarity once it has crossed the Alps into Italy.

stowaways

Many vagrants are passerines, a fact explained by their light weight and modest physical capabilities. Thus, American warblers come high on the list of vagrants encountered in Britain, Ireland, and France. Similarly, the Siberian leaf warblers are regular amongst unusual migrants. Nevertheless, passerines weighing a few ounces are not the only ones to succumb to strong winds or make navigational errors. Many ducks, waders, gulls, terns, and raptors are also vagrants. But while passerines often go astray because they are blown by the wind, other birds may reach our shores because they are strong flyers.

Very occasionally, birds from North America or Greenland reach Europe on ships crossing the Atlantic. Imagine a small bird, the size of a sparrow, lost over the ocean and fighting the wind for many hours. Exhausted, it flies lower, ever more dangerously close to the waves. Suddenly it catches sight of a ship - perhaps even a small one - and, summoning the last of its strength, alights on it. Many a lone sailor has told such stories, which are often very moving. Some birds make part or all of a crossing in this way, as long as food is available; there have been cases of such stowaways being fed by the crew and thus making the entire crossing effortlessly.

Above, top
Certain species are extremely rare vagrants to Europe. The sooty tern, which lives in tropical seas around the globe, has been seen in Europe fewer than 100 times during the 19th and 20th centuries.

Above, bottom
Marooned on an island on the far western fringe of Europe, this Isabelline wheatear has gone far off course. In normal circumstances it would have flown to Arabia or the Indian subcontinent for the winter.

Right
For some years now the ring-billed gull has been widening its range, reaching European coasts more frequently and even venturing inland. After spending the winter in this part of the world some of these birds return to their native North America.

<u>majestic at rest, and peerlessly elegant</u> when they take wing and settle into their ordered formations, common cranes are also touchingly faithful. Not only are pairs inseparable once formed, but family groups stay together far longer than in many other species.

The majestic flight of the crane

Cranes are equally faithful to their migration routes and the stops they make along them. In eastern societies the crane symbolises fidelity in love; in the West, we have long been fascinated by these birds of good omen that cross our lands twice a year, cleaving the sky with their long, undulating "V"s and alighting to recover their strength before flying on, tireless harbingers of the passing seasons. By slipping into their wake, we can follow each stage of their long journey which, autumn after autumn, takes these birds from the far north of Europe, where they nest, to the warm Mediterranean regions.

Top left and right
Adult common cranes have distinctive markings on the head, the crown has a bright red patch of bare skin which swells and becomes more prominent in the breeding season. Their dagger-like beaks are versatile, allowing them both to catch insects and grubs and to pick plants.

one day in Late summer ...

For some time now the days have been noticeably shorter in northern Europe. It is barely the end of August, yet the long days of June, when the sun disappeared only fleetingly below the horizon, are just a vague memory. Above the flooded Swedish peat bog, the morning mist takes longer to disperse, and there is a definite sharpness to the cool of the dawn. A few flurries of wet snow have already heralded the approaching end of summer.

Since April or May, the end of their spring migration, the common cranes have devoted themselves to breeding. After mating displays, nest building, and the month-long incubation period, the chicks were hatched. Now two and a half months old, the young cranes are the size of adults, and can only be distinguished from their parents by their plumage, which is reddish and a duller grey. Their heads also lack the red, white, and black marks of their parents. Now at last their huge, fresh-feathered wings, with a span of some 6½ feet, can be put to their proper use.

This morning the family groups are visibly restless. Their offspring fledged, and their pinions - long flight feathers on their wings - newly-grown after moulting, the cranes are ready for their long journey. Over the last few days they have begun to congregate. Their calls, like so many trumpet blasts, ring out across the sodden marsh, dotted with dwarf birch trees and the occasional conifer. There is no longer any doubt: it is time to leave. Once the mist has evaporated and the sun has finally warmed the air, the cranes feel that conditions are right to set off on their great flight. As they rise at last with powerful wingbeats, at least the adults among them probably know, if only dimly, that they are embarking on a voyage of several months, one which will take them thousands of miles away, to their wintering areas.

What of the young birds? They know only that they must follow their parents, and will stick tenaciously to them, constantly uttering their feeble cries as they fly, a peeping sound that seems incredible coming from such big birds. They probably do not know that they can rely on their parents' devoted help: adult cranes have been known to break their journey until one of their young, which had alighted in exhaustion, could continue.

setting a course for the Baltic

At the rate of a few hours' flight per day, the flocks of Swedish cranes head south - but not without allowing themselves breaks for, at the moment, time is not short. In Norway, too, the great southerly journey has begun. For cranes coming from western Scandinavia the traditional gathering sites are in southern Sweden and northern Germany. In the latter the island of Rügen, north-east of Rostock, and Lake Müritz, in Mecklenburg,

> " cranes fly very high, and arrange themselves in formation to travel; they form a rough isosceles triangle, as if to cut through the air more easily ... "
>
> **Buffon**

Above
On migration, and while wintering, common cranes seek the company of their fellows.

Right
Wonderfully streamlined to slip through the air with minimum effort, common cranes are capable of flapping flight, gliding, and soaring. With a following wind the migrants can reach about 68 miles per hour, but in general they travel much more slowly.

are host to tens of thousands of cranes. After some weeks spent replenishing their reserves, these vast flocks decide to begin the migration proper, from October to November. Once again, they will fly across a vast area of western Europe.

Leaving the Baltic region, the cranes cross northern Germany, heading for Luxembourg. They pass over the Ardennes and enter French air space.

When they make these journeys they settle into geometric formations: either long, undulating lines or "V" shapes, sometimes consisting of several hundred birds. When they are not flying too high they can easily be seen from the ground; however, clear skies sometimes allow them to go higher, and they can then travel above 3,300 feet - so high that they are almost invisible, despite their size. Thus great flocks of hundreds of birds can pass, crossing a huge span of sky in complete secrecy, as if slipping through a dream.

The champagne stage

Soon the great lakes of the Champagne region gleam in the sun, or the pale light of the moon, for these lovely travellers can travel by night as well as by day. At a safe height, so as to avoid the obstacles that might rise in their path in the darkness, and trusting the stars and their own mysterious

Above
On their journey from Scandinavia to Andalucia common cranes visit the same sites - such as the large lakes in Champagne and Brenne (above) - in large numbers. Changes in weather may determine how long they linger there.

sense of direction, the cranes fly resolutely onward, wing to wing. Each hears the reassuring sound of their neighbours' wingbeats. Frequent calls also help to keep the flock together. Audible several miles away, these cries also help several flocks to keep in contact during the migration.

Once above their target lake, the great birds abandon their flawless flight formation and begin to wheel in great circles, gradually coming down to earth. This is a crucial moment for the flock. It is the responsibility of the parents, who are suspicious by nature and from experience, to assess the safety of the site they have chosen. Far below, on the islands formed by the lake's dropping water level, recently-arrived cranes are settling down for the night and calling. Reassured, the new arrivals accept the invitation and soon alight, wings spread and legs outstretched to absorb the shock of landing. As soon as they are on the ground, families that were briefly separated regroup, and prepare to enjoy a well-earned rest after several hours and several hundred miles of continuous flying.

Left
The passage of a great flock of migrating common cranes is a spellbinding sight such as only nature can offer. Not just the eyes, but also the ears are thrilled, for cranes make full use of their voices. Their sonorous trumpeting calls - uttered especially as they approach a stopping place - carry for several miles and are unforgettable.

Right
After several hours' flying, the birds have at last reached their resting place. After first circling at height to assess the site's safety, the cranes break their ordered flying formation. Legs outstretched to absorb the jolt of landing, and wings arched to act as air brakes, these two cranes glide gracefully down to the ground.

The young cranes have just learned some precious information about migration: the course to follow, the stops to make, and the precautions to take before alighting. So strong are the ties between adult and young cranes that even on the return journey, after wintering, families can be seen travelling together, which gives the young birds a further opportunity to learn.

If weather conditions allow, some of the cranes will stay on, spending the winter in the Champagne region. This is a recent phenomenon, due to the creation of several large artificial water-storage lakes, which regulate the level of the River Seine. The cranes feel safe there at night, while the surrounding fields offer food. In the early morning they leave their night refuge to feed. Amid an astonishing chorus of trumpeting calls, and against the background of a milky, pink-tinged sunrise, the procession of birds lasts several hours when wintering or passage migrants are very numerous. From mid-afternoon until twilight, the same performance takes place, in reverse.

Beyond the Pyrenees

After a pause of anything from a few hours to several weeks, the cranes that have chosen not to stay in north-east France continue their journey south-westwards, crossing the country diagonally and stopping off when they need to in places such as the Brenne region – the "land of a thousand pools" – or the Landes, in the south-west of the country. Then comes the first real test for the young birds: crossing the Pyrenees, whose peaks, already snow-covered and often wreathed in ominous-looking cloud, now appear on the horizon. Fortunately, the cranes fly strongly and the ordeal does not discourage them: they climb and reach a pass, which makes their crossing easier. Once they are on the Spanish side, all that remains is for them to finish the last leg of their journey, pausing at their habitual resting places, until they reach their winter quarters in Extremadura and western Andalucia. There, at the end of their arduous journey, they can spend several months peacefully wandering around the *dehesa*. This wooded savannah is an ancient landscape consisting of pastures shaded by huge scattered evergreen oaks and ancient olive trees. The cranes will spend the winter feasting on acorns and olives that have fallen to the ground, without turning up their noses at insects, molluscs, worms, or grubs, for their tastes in food are catholic. And while in summer they are used to seeing reindeer or elk, here they rub shoulders with fighting bulls.

Some of the cranes go further, crossing the Mediterranean to winter in North Africa. Then, between February and April – depending on the latitude they have reached – the birds start to gather once again to prepare for the journey north-east, as they have done for tens of centuries. Under their wings will pass, in reverse order, the arid Spanish plains, the Pyrenees, France, Germany, the Baltic and, finally, the great peat bogs or tundra, where the last snows are melting.

Above
The great birds' breath can be seen in the freezing early morning air as they call noisily to each other. As in many gregarious species, vocal communication plays an important social role in cranes, and can be used to express alarm, aggression, desire for a mate, the need for group cohesion, or simply exuberance.

Right
During the winter months cranes wander through the peaceful landscapes of southern Spain, calmly feeding on acorns fallen from the evergreen oaks.

winter

If one were to draw up a League of migrants, there is no doubt that the long-distance travellers, especially those that go from one continent to another, would be top of the table.

short-haul flyers

Capable of covering thousands of miles, of pressing on determinedly night and day, and of crossing - not without effort - seas and deserts, they attract our admiration, as do all great travellers, whether human or animal.

However alongside these lords of the sky, which include some passerines, raptors, terns, and waders, is a huge number of generally less ambitious migrants, which are often content with rather shorter journeys - though not always, as we shall see. These "unknowns", these "also-rans", lacking the prestige of the great travellers such as geese or swallows, are known as partial migrants. However, the shortness of their journeys does not mean they are not worth studying, for they too have developed strategies for coping with new environments.

Preceding pages
Partial migrants that do not make long journeys may face tough winter conditions such as snow, which, for this robin, makes finding food difficult.

Top left
Squacco herons, which normally like warm conditions, are usually long-distance migrants, wintering south of the Sahara. However, some birds migrate shorter distances, and are content to winter in the Nile Delta.

Right
Although some goldcrests - despite weighing less than ⅓ ounce - are able to cover thousands of miles from northern to southern Europe, most of these little birds migrate only modest distances.

itchy feet

Many European birds qualify as partial migrants, and the numbers of individuals involved may be large. It should first be remembered that true resident birds are rather unusual. Even the most home-loving species, including most woodpeckers, nocturnal raptors such as barn or tawny owls, sparrows and larks, tend to wander, especially after breeding. However, these movements do not follow any particular direction, and primarily involve young birds in search of territory they can occupy without entering into conflict with their fellows - an aspect dealt with in greater detail in the last part of this book.

This fact about the small proportion of sedentary species helps to explain why there are so many partial migrants: a species is so described even if only some of its members migrate.

A glance at a bird field guide will make clear how partial migration works. Any reputable guide will have distribution maps with areas marked in three colours: the first shows the breeding range, the second the wintering range, and the third - often a mixture of the first two - the regions where the species is present all year round. When the three colours appear on the same distribution map, this indicates that the species in question is a partial migrant. The chiffchaff is a good example. The northern range of this little passerine - where it is only found in summer - comprises northern and central Europe; its southern range - where it only appears in winter - southern Spain, Greece, the Middle East, and North Africa. Between the two lies an area covering almost the whole of southern Europe, from east to west, where chiffchaffs are found in both winter and in summer. Insofar as the breeding and wintering ranges are not discrete, but on the contrary overlap to some extent, the chiffchaff can be described as a partial migrant.

Longer wings mean Longer journeys

Some families of birds include both partial and long-distance migrants. The willow and wood warblers, related to the chiffchaff mentioned above, are both long-distance migrants. After breeding, they leave Europe altogether for Africa, crossing the Sahara. Interestingly, the chiffchaff's wings are shorter than those of the otherwise almost identical willow warbler; there is a clear correlation here between the tool and its performance.

> " The Last swallows, caught unawares by the cold, loop listlessly around the flocks in the meadows, or skim the walls of houses with heavy wingbeats, searching in vain for flying insects. "
>
> **Jacques Delamain**

Left
As winter draws near, the buzzard may behave in a variety of ways. Some birds remain almost sedentary or travel short distances, while others may cover several thousand miles - flying, for example, from Sweden to Morocco.

Right
The brambling's migration does not take it beyond Europe's borders. Shortage of food may force it to leave the region where it is wintering. It is very fond of beech mast, on which it feeds with gusto.

Another example is the garganey, which leaves Europe at the end of summer and heads for Africa, while the teal is content to stay within western Europe and the Mediterranean fringe, among other places.

To give an even better idea of partial migration it should be pointed out that certain species are regarded as true migrants within Europe, but classed as partial migrants over their whole range. The purple heron, for example, is only a migrant within Europe: present in the breeding season and totally absent in winter. However, in a broader context the species is a partial migrant, in that it can be seen all the year round in parts of the Middle East and in Asia. To sum up: within a given partial migrant species, some individuals make short journeys, while others travel as far as true migrants.

move away or stay at home?

There are partial migrants in almost all families of birds. To name but a few: among herons, grey heron; among waders, lapwing and common snipe; among gulls, herring gull and great black-backed gull; all belong to this category. A large proportion of passerines are partial migrants too. Blackbirds and thrushes, some buntings, many finches (goldfinch, linnet, chaffinch, brambling, redpoll, siskin) are for the most part short-distance travellers or, at any rate, do not have discrete

breeding and wintering ranges. It is easy to establish whether a species is a true migrant since it alternately disappears and reappears, as do swallows or reed and sedge warblers. Similarly, in the depth of winter it would be pointless to look for a swift in our latitudes, whereas, from April to August, it is almost impossible to gaze up at the sky without seeing at least one of these sickle-winged birds. With partial migrants, however, it is harder to establish their movements without tracking individual birds. Such species may be present in a given region all year round, but this may involve different individuals.

Far left
The grey heron's migration generally ranges from a few tens of miles to a few hundred, but some, blown off course, have been known to cross the Atlantic.

Left
Capable of exploring the thinnest twigs thanks to its acrobatic skills, the redpoll hardly ventures beyond central Europe, whether migrating or wintering. It is a nomadic bird, moving about according to local food availability and weather conditions.

Top right
In western and southern Europe the fieldfare is a migrant, spending the winter there while still being affected by changes in weather conditions. If snow sets in, these large thrushes are forced to move further in search of milder conditions - an "escape migration". The bird pictured has drawn its left leg into the warmth of its belly feathers.

Below right
The end of the common snipe's long beak is extremely sensitive, allowing it to find worms buried in soft mud. If a freeze sets in, probing the mud becomes impossible, and the snipe must fly to more hospitable regions. However, these birds are not short of initiative and, if there is a sudden cold snap followed by a snowfall, they know how to explore the sheltered places under trees in search of invertebrates.

the robin in my garden

Ringing programmes have gradually allowed a picture of the movements of various species to be built up. One example will suffice to illustrate the complexity involved. Someone whose garden, in lowland western Europe, is visited by one or more robins all year round might imagine that the same bird appears every time, and people sometimes talk fondly of "my" robin. However, at the risk of disappointing such sentimental gardeners, it is possible, even likely, that several birds have passed through their patch.

Let us look at the situation chronologically, starting, for convenience, in spring. For some days, following an improvement in the weather, a robin has been singing at the top of its voice, proclaiming control of its small territory. Travelling by night, it has come from southern France or Spain

where it spent the winter, arriving as soon as that season was over. When autumn comes, it may return to warmer regions. For some weeks its absence will go unnoticed, because one or several robins from northern Europe will come to take its place, sometimes only for one day. Finally, one of these, deciding that it has travelled far enough south, may decide to settle here for the winter. If it survives it without starving or being caught by a cat, the following spring it may make way for a robin that may nest, and the cycle can be repeated. It is equally likely that if the same wintering robin suffers a mishap it will be replaced by another, looking for winter territory. Everything depends on how attractive a given site is. If it offers a good habitat, it is unlikely to remain unoccupied for long, especially since in territorial species there are usually more individuals than territories, so that the less competitive birds have to wait in peripheral, sub-optimal areas until one falls vacant.

Above
If harsh weather persists in winter, only the most resourceful robins can survive. Armed with experience, adult birds are more likely to cope in tough conditions.

Right
The hawfinch, reminiscent of a small parrot, is a partial migrant in Europe. Some are virtually sedentary, while others travel short distances. Most Russian hawfinches however cannot withstand the rigours of the winter there, and are true migrants.

At the end of their autumn journey migrants find - or in some cases revisit - the region where they will spend the winter.

winter quarters

As we have seen, some migrants actually spend only a limited amount of time in their wintering area proper, since the southward and northward journeys can take so long.

Top left
In summer the monotonous cooing of the turtle dove can be heard in the lowland pastures, copses, and woods over much of Europe. Come winter, this sensitive bird haunts the Sahel, from Senegal to Ethiopia, frequenting dry grassland, fields and paddy fields, where it feeds on seeds and grain.

Right
The black-winged stilt is heavily dependent on water for nesting, migrating, or wintering. Depending on the latitude and time of year, it dips its long legs and sharp bill into pools, lagoons, or paddyfields.

From one world to another

Depending on the species, wintering ranges may be within Europe itself – in which case we speak of short- or medium-range migrants – or on another continent: usually Africa, but also the Middle East or the Indian subcontinent. These transcontinental birds are for the most part long-range migrants, capable of covering several thousand miles. However, a bird can perfectly well pass from one continent to another without having to travel a long distance. A stone curlew – a nocturnal wader of dry country, with large yellow eyes – nesting in Spain and spending the winter in Morocco is thus just as much a transcontinental migrant as a ⅓-ounce willow warbler

that covers the astonishing distance from the Baltic coast to southern Africa.

Whatever the length of their journey, birds that can be described as "true migrants" are those that breed in a given region and winter in another completely separate one, the two perhaps several thousands of miles apart. The dainty red-footed falcon, with its handsome slate-grey plumage and brightly coloured legs, nests in central Europe and winters in the African savannah south of the equator. The nightingale, whose flutey melodies lend magic to spring nights over much of Europe, winters in tropical Africa. Finally, the wryneck – a bird with bark-coloured plumage and an atypical member of the woodpecker family – is found over most of Europe during the breeding season but then leaves for Africa, even south of the Sahara.

variations

Migrants whose breeding and wintering ranges overlap even a little are known as partial migrants. These distinctions between different categories of migratory birds are described only briefly here, to clarify them and the terminology applied to them. But in fact matters are a good deal more complex if we consider the different "populations" of birds belonging to the same species, but inhabiting different regions.

Far left
After breeding, the ruff travels in the space of a few weeks from the Finnish marshes, to wet grassland in the Netherlands, to sun-drenched African swamps. It can be found either side of the equator and as far as South Africa.

Above right
Unlike most wetland warblers, which are long-distance migrants, the fan-tailed warbler has opted firmly for the sedentary life. However, if the winter is particularly harsh, it pays for this by dying in large numbers, which are only replenished after several years.

Right
After a summer spent wandering perhaps as far as the Channel, the Mediterranean shearwater returns to its namesake sea, where it spends the winter months moving easily from place to place, thanks to its excellent flying capabilities, even in fierce winds.

Following pages
No doubt these common cranes, their shadows lengthening in the setting sun on a migration stop or in their winter quarters, have met up with their fellows on alighting.

Thus a stonechat, for example, can be considered a long-distance migrant if it nests in Germany and winters in North Africa, or a short-distance one if it breeds in Belgium and spends the winter in southern France - and almost sedentary if it both nests and winters in Brittany, only moving from the interior to the coast.

A change of scenery

One of the most surprising aspects of migration is that it leads birds to frequent, different habitats during the breeding season and in winter. Certainly a blackbird nesting in a Belgian garden and wintering in a French park is not, strictly speaking, changing its surroundings. But what of a dunlin, a small wader that breeds, for example, in the Swedish tundra and winters on the mudflats of the Golfe du Morbihan in southern Brittany?

There are other examples of this wonderful ability to adapt to completely different environments. Bewick's and whooper swans that in summer frequent lakes in the Arctic tundra, comb the meadows and stubble fields of the Netherlands in winter. The swallow that built its nest in the rafters of a stable in a Bavarian village and spent the summer chasing insects above pastures filled with placid cows, finds itself a few weeks later flying over the African savannah, dotted with acacias and populated by herds of zebras and prides of lions. The cuckoo hatched in the nest of a reed warbler deep in a Danish reedbed will go and spend the winter in the heart of the Gabon forest. Just as extreme, perhaps more so, is the change experienced by phalaropes with the seasons. These delicate waders, which nest in the Arctic tundra, are unusual in that the females are more brightly coloured than the males, and leave them the job of incubating the eggs and bringing up the young. After mating and laying, female red-necked phalaropes head south to winter in the Gulf of Oman, while female grey phalaropes go to the Atlantic ocean off western and southern Africa. The males and juveniles follow later. All will spend the winter months at sea, feeding on tiny invertebrates picked from the surface.

It should be pointed out that many species winter in the open sea. Some of them are completely pelagic, only visiting land to nest briefly on sea cliffs. Auks - guillemots, little auks, and especially puffins with their bright, parrot-like bills - gannets, and divers, with their marvellously streamlined bodies, spend the winter months and much of the rest of the year out at sea. Skuas, related to gulls, together with shearwaters and petrels, cousins of the albatrosses, do not confine themselves to any one stretch of sea, but embark on long voyages over the oceans, which take some of them to waters off the coasts of South America or western Africa.

Above
Many European migrants leave to spend the winter under the tropical sun. There, they encounter conditions radically different from those they are used to in the breeding season, such as rugged, arid country in Niger (top: a watering hole in the Aïr), or wetlands in India (centre: the Bharatpur marshes in the north-west of the country) and eastern Africa (bottom: the Lake Bogoria area).

Right
Like a handful of crosses tossed up into the sky, garganey fly swiftly over the harsh mountains of the Middle East, producing a surprising contrast between these water birds and a desert habitat. This pretty little duck, its wings marked with blue, is a long-distance migrant, flying from the marshes of Europe to wetlands surrounded by the grasslands and arid areas of Africa, from Senegal to the Nile.

" ... their flying skeins in the mornings of the world "

Bernard Lorraine

Different times, different habits

It is not just the scenery that changes for migratory birds in winter: some species change their behaviour too. The wood sandpiper, for example, likes to perch in trees when it is nesting in the damp taiga of northern Europe, but in winter it loses this habit and keeps almost exclusively to the ground. Similarly, birds that breed in isolated pairs suddenly acquire a taste for the company of their fellows during the winter. Larks or thrushes, for example, sometimes form large flocks in winter, but during the breeding season they fiercely defend their individual territories, however small.

The same is true of waders such as lapwings, bar-tailed godwits, ruffs, and golden plovers. Although they nest apart, in winter they form flocks numbering up to several thousand birds, closely massed together. Many ducks also become highly gregarious throughout the winter, whether these be sea ducks such as eiders or scoters, dabbling ducks such as the wigeon, teal or pintail, or diving ducks such as the scaup, tufted duck and pochard. But pride of place - at least in Europe - goes to a small Scandinavian passerine: the brambling. After nesting in the northern forests, it leaves in autumn and heads for central Europe. There it sometimes

forms very large flocks, which in some years become truly breathtaking. By day, the wintering bramblings scatter in all directions to feed on seeds of various kinds. In the evening they gather into roosts; in the winter of 1999-2000, one of these numbered some 20 million birds.

Above
Winter haunts are sometimes the scene of some spectacular gatherings. In the soft evening light, hundreds of cormorants mass to roost in the trees.

Right
Wintering ducks are sometimes so numerous on lakes, where they feel safe, that they almost cover the surface of the water. Completely confident, these pochards drift off to sleep. Despite this apparent quiet, the slightest alarm will cause the entire flock to wake in an instant and take to the air, making the water boil.

Migratory birds that reach southern Europe or press further south into Africa find refuge, for a time, from the rigours of winter. This is not so for those that winter in Europe, particularly in central Europe. It is rare for snow or severe cold not to render these hitherto welcoming regions inhospitable.

Fleeing bad weather

When this happens the migrants must continue their journey, usually towards the west, where the ocean keeps temperatures higher, or to the south. This is called "escape migration". Localised cold snaps lead birds to leave the affected area, whereas longer, more severe, or more extensive cold spells give rise to a massive, often spectacular, exodus as millions of birds are forced to flee the cold. Naturally, migratory birds are not the only ones affected; resident birds also suffer - perhaps more so, as they are less inclined to travel. These species suffer the heaviest casualties in winter, sometimes to the point that their populations need several years to replenish their numbers.

Top left
Chaffinches do not necessarily remain in their winter haunts but react to changes in weather by moving about periodically.

Right
If snow stays on the ground, ground-feeding species must leave for regions with better weather. These geese cannot graze on grass as they usually do and, deprived of food, must move on to survive.

when the snow falls ...

Although the outcome is the same – enforced departure – not all species move on for the same reasons. When the snow is heavy and long-lying, the birds affected are those that feed exclusively or mostly on the ground, in areas such as meadows or fields. Passerines affected include skylarks, pipits, wagtails, buntings and – to a lesser extent, since they can feed in trees or bushes – thrushes and seed-eating finches such as chaffinches, bramblings and linnets. Among waders, those that stay inland in winter, such as lapwings and golden plovers, are likely to be forced to move on. And a broad range of other birds face difficulties: wild geese can no longer graze; raptors such as buzzards and kestrels have problems finding field voles under the snow, and woodpigeons are deprived of a large part of their diet. Similarly in gardens snow can affect species such as blackbirds, hedge sparrows or the familiar robins, sometimes severely.

... or the ground freezes hard

Even if snow has not fallen, as soon as cold, especially bitter cold, sets in, other migrants suffer. The first are those that frequent water. Ponds are the first to freeze over, followed by larger pools, and finally lakes. As each does so, water birds are forced to move on. Finally, when no open water is available, the birds must resort to rivers, whose flowing water freezes less easily. If even the watercourses freeze over, there is nothing for it but to flee southwards, or to the sea – assuming the birds are physically up to the journey and that their ecological needs can be met by these new regions. Here, too, migrant species are better equipped, in theory, to deal with such a critical situation. The long list of water birds threatened by frozen fresh water naturally includes those that feed on fish – goosanders, cormorants, herons, and kingfishers – as well as those that are wholly or partly vegetarian: mallards, teal, diving ducks – such as the scaup, tufted duck and pochard – coots, and moorhens. Nor are the strictly aquatic species the only ones to suffer during a freeze. All birds that depend on wet areas and find their animal or vegetable food in the earth, in humus, or in mud, are faced with ground too hard for their beaks to penetrate. In the forests, woodcocks; on the plains, lapwings or golden plovers; in the marshes, common snipe; and in the meadows, jack snipe: all these birds are suddenly deprived of their food. For these too, the only hope is to migrate to regions where there is no freeze – at least not yet.

Above, top
Gripped by ice, the banks of the lake have been deserted by birds. The cold winter sun will not be enough to bring a thaw, and conditions will remain harsh here for some time.

Above, bottom
Seed-eating passerines, such as this yellowhammer, can survive in snowy conditions by slipping under the grass to pick up fallen seeds – a precious resource in this time of scarcity.

Right
Black-headed gulls depend heavily on wet areas. When a freeze sets in they have to move en masse to watercourses – where some of them gather even in milder weather.

Following pages
These barnacle geese appear disconcerted by the snow covering their wintering grounds; in fact this Arctic species is familiar with such conditions.

<u>mountains are the scene</u> of a unique migratory phenomenon. At the end of every summer - sometimes earlier, in July or August, depending on the weather - those birds adapted to nesting at high altitudes find the conditions deteriorating.

mountain migrants

Icy nights, freezing rain, and snow flurries force most mountain birds to leave the high peaks. These specialists in altitudinal migration start to head towards the valleys. The species involved are mostly passerines such as water pipits, whinchats, tits and black redstarts. But raptors such as the golden eagle and peregrine are also involved, as are grouse and capercaillies.

Top left
Although the black grouse tends to be faithful to its haunts, it may move in winter to find more plentiful food. Females do so much more than males, though it is not known why.

Right
Known as the "snow partridge" to some mountain dwellers, the ptarmigan in its immaculate winter plumage is perfectly adapted to its life in the high mountains. Nevertheless, come winter, it too must make some concessions and descend a little, if it is to continue foraging for its diet of buds and pine needles.

camouflage clothing

The birds best adapted to high altitudes, such as the white-winged snow finch, alpine accentor, and wallcreeper - whose wide wings marked with red recall those of a large butterfly - are forced to descend a few hundred feet. Some of these birds readily go further on to the plains, sometimes covering several hundred miles in terrain quite unlike their usual surroundings. Though better equipped to cope with severe weather, ptarmigan and choughs also have to retreat somewhat when faced with snow and ice.

Ptarmigan are remarkably well adapted to their winter habitat. Their plumage, which is brown in summer, gradually turns to white in winter, ensuring effective camouflage. Their legs and feet are also covered with feathers, which has the double advantage of protecting them from cold and acting as snow shoes. Furthermore, ptarmigan, like their cousin, the black grouse, have mastered the technique of igloo building, digging themselves snow holes in which to spend the night.

Alpine aerobatics

Alpine choughs are not to be outdone. These little members of the crow family perform stunningly agile aerobatics and hold records for high flying, being able to ascend to more than

26,000 feet without difficulty. By day they go down into the valleys - where they have learned, notably, to take advantage of the feeding opportunities at ski resorts - and then return to spend the night high up, where they shelter in rock crevices, unconcerned by the ferocious drops in temperature in this harsh environment.

When spring comes, as soon as weather permits, those species that fled the high mountains promptly return.

Left
The alpine chough - an astonishing and masterly flier - tends to descend a little in winter, even though it is remarkably well equipped to cope with the harsh conditions at altitude.

Right
Although it has learned to take advantage of the wealth of food scraps around ski centres, the alpine accentor readily leaves the high peaks in winter to take shelter on lower hills.

Migrants' movements seem to be choreographed with such precision that the dates on which they pass certain points, the arrival of wintering birds and the return of breeding ones can be predicted, sometimes with astonishing accuracy.

invaders

However, in some years "nordic" species make completely anomalous movements that follow neither their usual calendar nor their habitual routes.

These mass departures, from which birds often do not return, always involve species that come from far northern latitudes such as Scandinavia or Siberia, and can involve vast numbers of birds. They are a true survival mechanism: driven by hunger, the birds leave early for an unknown destination, where they hope to find some nourishment.

Top left
Some years the nutcracker, recognisable by its long, slender bill, leaves its cold native Siberia to disperse over much of Europe. These mass movements are driven by the need for food, which has become desperately scarce further north.

Right
The waxwing, a true winged jewel, much sought after by European ornithologists, is forced in some years to leave its normal range to search for its favourite food, berries.

Following pages
In a picture reminiscent of a Japanese painting, a flock of wintering waxwings feed on rowan berries.

In search of seeds

In autumn, and to a lesser extent in winter, the vast northern forests contain abundant food for bird species that feed on all manner of fruit, berries and seeds. The conifers of the taiga - pine, fir and spruce - are covered with quantities of cones that contain plenty of seeds. Birches also offer seeds - tiny, but in large numbers, and rowans provide handsome, glossy red berries.

Conifer seeds are the staple diet of many species, such as the nutcracker and the crossbill, which is reminiscent of a small parrot. As its name suggests, this bird has mandibles whose hooked ends overlap - a perfect adaptation for extracting conifer seeds. Each species of crossbill prefers a different type of seed, thus reducing competition. Birch seeds attract various tits such as the coal tit, as well as siskins and redpolls - tiny acrobats that can dangle upside down at the end of the slenderest twig. Finally, rowans are a feast for those lovely, colourful passerines with their unique crests - waxwings. In normal conditions all these birds find enough food in the cold regions they inhabit, even in winter. Unfortunately, however, nature sometimes fails them. Thus one or other species of tree may produce an unusually poor crop in a given year - on a fairly regular cycle. Deprived of their normal rations, nutcrackers, crossbills and waxwings have to flee southwards to seek supplies. This sets large numbers of birds on the move in autumn and early winter.

A shortage of rodents

Although on nothing like the scale of the invasions of seed-eating species, birds of prey are sometimes forced to leave northern Europe and travel south in winter because their usual prey - small rodents such as voles and lemmings - is in short supply. This phenomenon is cyclic and happens roughly every four years, though the reasons are unknown. When it does, snowy owls - magnificent nocturnal raptors with white plumage - rough-legged buzzards and great grey owls leave the tundra and the great coniferous forests, sometimes turning up in central Europe, France and Germany, and even as far south as Spain or Italy.

Above
The crossbills that live in the conifers of northern and eastern Europe sometimes move about in large numbers, starting in summer. Their colourful, wandering flocks may then be seen throughout the winter.

Right
Nutcrackers that live in the Alps do not invade other areas as do their Siberian cousins. When birds from the Alps appear in lowland forests this is due to nomadism.

spring

Their winter sojourn over, migratory birds must think about setting off for their breeding areas once again. For European birds, this means heading north.

The flight home

Migrants that escaped the many dangers of the southward journey must face more on the return leg. Their motivation for doing so is a powerful one: perpetuation of their species.

Preceding pages
The blue sky is reflected in the calm water, the water lily petals open, the warm days are back - and with them this black-necked grebe, showing its ruby-coloured eye.

Top left
At the end of its spring migration the male Montagu's harrier returns to the great open spaces where it raised its young last year.

Right
Journeying north from Africa in long stages, the wheatear is an early migrant, returning to Europe in early spring. It likes to perch on a hillock or rock, from where, ever alert, it spies insects on which it pounces in a brief flight, before fluttering back to its vantage point.

returning to their roots

Earlier in this book we saw how the movement of birds between their northern nesting areas to their southern wintering zones in Africa or even the Antarctic may have arisen. However, although researchers are coming up with increasingly well-documented explanations, one question remains: by what strange mechanism can a bird that has been wintering in southern Europe or Africa for several months "feel", after a certain length of time, that it must leave and make the reverse journey? As in the period before the bird's first departure, internal changes initiated by its endocrine glands prepare it for the return migration. The main change, linked to the bird's reproductive system, brings with it a growing compulsion to breed.

It is understandable that a European bird may not be able to breed in its winter haunts which, as we know, may be very

different from those where it breeds. What has the Lapland tundra, in the far north of Sweden, in common with a sun-baked sandbank off the coast of Mauritania? There is little similarity between the lush, wooded countryside of Normandy, England or Denmark and the vast African savannah, dotted with acacias. However, it is harder to understand why a bird chooses not to nest in its winter haunts if the conditions there are closer to those where it habitually breeds. Force of habit, deeply rooted in its genetic make-up, is certainly an essential factor. It should be added that, at the end of the winter, southern Europe contains large numbers of birds of species that cannot travel further north. If the birds that had spent the winter there chose to stay put, there would be severe overcrowding. All this is, of course, speculation, since for the moment there are no full or satisfactory explanations.

However, where seabirds are concerned - especially those that winter far out to sea - the problem is a different one. These pelagic species could not breed if they did not return to land. Thus kittiwakes, puffins, petrels and shearwaters are forced to leave the oceans where they spend the winter to return to their nesting colonies on steep sea cliffs, or on islets battered by the waves.

Above
Snow-covered and in twilight for many months of the year, the Arctic latitudes enjoy a late but bountiful spring. Marshes, peat bogs and tundra are once again home to a multitude of birds such as great (centre) and Arctic (right) skuas. So close to the North Pole, these species have only a few weeks in which to breed.

Right
Free of ice at last, Arctic lakes are home to several species of diver.

moult or migrate - birds must choose

During the winter months migratory birds have been doing two things: one continuously - feeding - and the other at a specified time - moulting. The two are closely linked, since a bird must be in good physical condition - and therefore properly nourished - in order to moult, which is a physiologically demanding process.

All birds moult at some time - replacing their old feathers and growing new ones - because feathers eventually wear out and lose their strength. Friction - against each other, but also against branches or rocks - is the chief factor in this wear, especially at nesting time when birds have to come and go from the nest carrying food for their young. Exposure to light is another cause of wear and pale feathers, or the paler parts of dark feathers, which contain less of the dark pigment melanin, tend to wear out faster. Accordingly, after a period that ranges from a few weeks to a few months, old feathers are shed and are replaced by new ones. Migratory birds generally moult twice a year. One moult - known as a "partial moult" - involves the feathers that cover the head and body; the other - known as the "complete moult" - involves the progressive replacement of all the bird's plumage. Since the complete moult also involves the flight feathers - that is, the primary feathers of the wings that drive the bird through the air, and the tail feathers, that act as a rudder - migratory birds cannot go through it at migration time, which is just when they need all their powers of flight. There are therefore only two solutions: to moult before migrating, or to do so while wintering. Thus the golden oriole, a magnificent passerine (the male is bright yellow) slightly bigger than a blackbird, undergoes a partial moult during the summer, before its autumn migration to eastern or southern Africa. There it will benefit from a complete moult between November and February, before embarking on the return journey to its European breeding grounds. The icterine warbler, a green and yellow warbler, also undergoes a complete moult in winter, just before its spring migration, but may go through its partial moult either in late summer, in Europe, or in early winter, in Africa. In all cases, migration and moulting are mutually exclusive.

These are the results of evolution and natural selection; but there are, naturally, exceptions to this general picture. Thus a very small number of species, such as the willow warbler, moult completely twice a year, while others, such as the nightingale and redstart - both long-distance migrants - only moult once, of necessity completely. Species whose members do not all migrate long distances - such as the starling, brambling and jackdaw - commonly moult just once a year.

Above, from top
Sky-blue or golden, the feathers of the kingfisher, jay, and goldfinch are revitalised with the coming of spring.

Right
The sap swells the silvery willow buds, and the soft pink of the bullfinch foreshadows the explosion of colour that reawakened nature will bring.

Certain migrant species even go through an "interrupted" moult: they start to replace their primary feathers while in their breeding grounds and then, when the time comes to leave, stop moulting. They then resume their moult during the winter. This can happen with sand martins, which change their primaries before migrating and finish their moult afterwards.

To give an idea of how long a moult can take, a pied wagtail - a small, black and white, long-tailed passerine - needs two and a half months to renew all its feathers.

Hurrying home

In the autumn, most migratory birds take their time, so to speak. Many set off while the weather is still good and enough food still available. Moreover, the adults do not face the heavy

task of nesting. As for the juveniles, they are in no hurry, either because they are cautious or because they are inexperienced. In short, there is a certain feeling of nonchalance around, and that it's fine to dawdle, especially if the mild weather holds out.

In spring, the factors involved are completely different. Potential breeding birds must hurry back to their nesting grounds, as the urge to ensure the species continues is strong. This is even more the case with birds flying back to nest in far northern latitudes: they have only a short window of time during which the taiga and tundra, freed of their snowy covering, become hospitable. Often the spring migrants' journey takes only two-thirds as long as the autumn one - sometimes only half as long. A small passerine such as the wood warbler, which breeds especially in our beech woods, takes two months to make its autumn migration, but only one month for its spring journey. This speed on the spring migration is even more noticeable if, rather than looking at all species, we look at only one - or better still an individual bird. While the migration of all species taken together takes place over several months, a given species or an individual member of that species will only take a small part of that time.

However, many birds, as in autumn, do not travel far at any one time and prefer to spread out their effort. A warbler, for example, may take a little under 100 days to travel the 3,100 miles to its nesting site. In theory, this means a journey of only some 31 miles per day - or rather per 24-hour period, for many warblers travel by night. Flying at the rate of some 19 miles per hour, the warbler need only spend two hours per day travelling.

Above
The measured pace of autumn migration is gone. In spring, driven by the vital need to reproduce, common cranes hurry back to the far north.

Right
The rock thrush has rediscovered its stony territory, which it brightens with its vivid colours. This bird is relaxed, standing on one leg.

These are only averages, to give a rough idea: in practice migrants do not necessarily travel every day. On the contrary, they may stop at a favourable resting place or remain grounded by poor conditions such as fog or persistent heavy rain. This being so, the length of individual legs and the time spent flying need to be revised upwards.

Some birds, especially those that are strong flyers, migrate much more rapidly than the average or, at any rate, cover longer distances in each stage. Waders, falcons, terns and certain passerines, among others, travel at considerable speed. Wheatears, for example, can cover 280 miles in one night, and there is a recorded case of a turnstone that covered 510 miles in 25 hours. A red-backed shrike can cover 310 miles in just one night on migration, flying for 10 hours at an average of 31 miles per an hour – a substantial feat that necessitates considerable rest afterwards. The bird will not repeat its exploit immediately, but will grant itself two days (which it will spend feeding) and one night's rest before flying on the following night.

Backtracking

Like the autumn migration, the spring one can run into bad weather: mostly in Europe, and usually in the centre and north of that continent - although spring can be especially cool and wet in Andalucia, for example, through which a great many migrants pass on the way back from Africa. Strong and unfavourable winds, downpours, late snowfalls, hailstorms and low temperatures can seriously affect migrating birds. However, too much should not be made of this: these "little balls of feathers" are less fragile than they might at first appear. For many thousands of years migrants have faced more or less serious weather almost every spring, yet the time of their arrival in Europe has not changed much. Despite this healthy resilience, if conditions become really tough the migrants may temporarily retrace their steps, a phenomenon known as "retromigration". Normally this allows them to find shelter but, if the weather is equally bad further south, their situation can become critical.

Trapped by the Light

The dangers migrants face include man-made ones such as lighthouses and, to a lesser extent, large, brightly-lit buildings. Powerful lights have an irresistible attraction for night migrants, which find themselves literally drawn like iron filings to a magnet. The birds, especially small passerines, fly in circles around the light source and end up exhausted or, completely blinded, crash into the lighthouse's window panes or into the building itself.

Above
These greylag geese, too, have heard the call to return. Night and day, they stream northwards.

Right
With its feathered warrior's mask and bright gaudy colours, the bee-eater brings a little bit of Africa to Europe.

Following pages
After flying over the Sahara and the Mediterranean by night, the wryneck, with its bark- and lichen-like plumage, returns to the parks, old orchards and woods of Europe.

Migrants are sometimes even attracted to the flames produced by gas flared off from oil fields, especially on offshore platforms such as those in the North Sea. In the early morning the platform decks are occasionally strewn with small corpses: passerines have flown into the metal structure or been burned by the huge flames. Though less dramatic, skyscrapers and other glass towers that remain lit all night are also migrant traps. Studies are under way to establish the true extent of this phenomenon. At the very least, such buildings should be left dark during the migration season.

Thoughtless poachers

Every autumn some migrant birds fall victim to hunters, but at least this activity is regulated by law. This is not so in spring. Most spring migration takes place after the shooting season has closed – which is when poaching occurs. What is most appalling is that the victims are birds that have survived postnuptial migration and wintering, and are preparing to breed: in many species, breeding pairs have already been formed during the winter. The most notorious blackspots for this activity are Mediterranean islands such as Malta and Cyprus, where certain species, especially passerines and raptors, face near-extermination. The destruction is carried out on a large scale, either with guns or with traps such as limed twigs. The birds, exhausted by a long night flight over the sea, are caught early in the morning when they come to land in search of a welcome perch. This is motivated by the desire for financial gain, or done simply for fun. There are problems, too, in Italy and France. In the former there is a dismal "tradition" of shooting at honey buzzards on spring migration. The slow flight of these handsome raptors makes them an easy target and when they finally reach the Italian coast, unscrupulous "marksmen" shoot them down like fairground targets.

In France, turtle doves returning from Africa are "hunted" in May in the Médoc, in defiance of European legislation and common sense. After leaving the Sahel, where they have wintered, the turtle doves, with their nervous flight, bravely cross the Sahara and the Mediterranean. Then comes Spain, and the crossing of the western Pyrenees, but before reaching north-west France or Britain they must follow the Atlantic coast. Eventually, with the Atlantic on their left and the wide Gironde estuary on their right, the doves seek to fly over land as far as possible: the only way is to aim for the Grave headland. Here death awaits them: the fusillade is let fly, and a hail of lead shot pierces their delicate pink plumage. These turtle doves, which surmounted so many obstacles, will not nest this spring. They lie instead on a bed of dry needles under the pines. In killing these potential breeding birds, poachers threaten their pitiful quarry with extinction: the species has declined sharply in western Europe.

Above
For the turtle dove there is no "merry month of May". On its return journey it needs a great deal of luck to escape the murderous gunshots that await it, illegally, in many European countries.

Right
The black-eared wheatear's spring journey is sometimes brutally cut short. This handsome bird is one of those awaited by a multitude of traps set by unscrupulous poachers on Mediterranean islands, such as Malta and Cyprus.

After weeks or months of travelling, the migrants have reached the end of their spring journey - but their troubles are not over.

Display time

Rest is the last thing on their minds. Breeding - an urgent, indeed fundamental, task - awaits them and will grant them no respite. On all sides, birds are preparing to breed, but first a male must find a suitable territory, defend it from rivals and attract a like-minded female.

Top left
A pair of fulmars share a tender moment at their nest. In many species, mutual preening is one of the behaviour patterns that reinforces ties between the two partners.

Right
Common cranes pair for life and remain very close. Mating displays start as early as winter, and during the spring migration. With ruffled plumage, necks outstretched, and making a thunderous clamour, the pair try to outdo each other in reaffirming their mutual attachment.

out to conquer territory

The problems faced by migrants arriving in their breeding areas vary from species to species, but their prime concern is the same: to find the space they need to build a nest and raise their young. Species that nest colonially, such as terns, gulls, spoonbills, certain herons and guillemots, return to their traditional nesting sites - unless these happen to have radically changed. The only space each pair need occupy is the bare minimum required for a nest. The distance between neighbouring nests often corresponds to the length of their occupants' necks - out of prudence, they stay just out of reach of pecking bills! As soon as the breeding birds return to their sandbank, marsh or cliff, the colony is established, amid a few ritualised squabbles and a noisy chorus of angry or indignant calls. First time nesters are pushed out by senior pairs, which keep the best spots for themselves. The newcomers make do with nesting at the edge of the colony, and wait for space to become available the following spring.

Non-colonial nesters must also find, first and foremost, a suitable nesting site, but in general they are less tied to a particular location. This does not mean, however, that they settle down anywhere, at random. On the contrary: ringing has shown that certain birds remain faithful to a particular area and may be remarkably consistent in their choices. This fidelity had been suspected, or established in the case of certain species with particular habitat requirements - cliff-nesting raptors such as the Egyptian vulture, and many water birds such as crakes, waders and ducks. Since these birds have a more restricted choice, it is natural that they should tend to seek out the same sites, which they know will suit them, year after year. But attachment to a nesting site has also been observed among passerines, which are less specialised and which could, it might be assumed, easily find many suitable sites. It is nevertheless astonishing to think that a tiny warbler with a minute brain can - after its first migration, a winter spent in Africa, and a spring return migration - select the same wood in northern Europe where it nested the previous spring! This is an extreme example, but birds quite commonly choose at least the same general area each spring.

Of course migrants, especially those without specific habitat requirements, sometimes change the area where they nest. In such cases any combination is possible: thus a bird might nest in western Europe, winter in Africa and, the following spring settle to breed in an east European country. Many studies of swallows have shown all manner of variations. These passerines may return to exactly the same place, be content just with the same area, or visit an entirely different one. It has even been noticed that swallows may, in the course of a single spring, spend time first in one country and then, towards the end of the season, move to another. A mystery indeed!

Above
The southern great grey shrike has an extraordinarily wide vocabulary: sharp or flute-like cries, whistles, bubbling calls, and an imaginative song that combines many sounds with mimicry of other birds' songs.

Right
The corn bunting is a mediocre performer, tirelessly repeating the same jangling song.

Home guards

Once a territory has been chosen, it is time to proclaim ownership. For many birds this is done essentially by means of various sounds, the most important of which is song. Passerines have become masters at this, as we can appreciate in our own garden by listening to the song of a blackcap, song thrush or robin. Generally speaking, the male is the first to arrive on the territory; almost immediately, he starts to sing ceaselessly. Nightingales, warblers, and wheatears (among many others) behave in this way. Patrolling his territory from one singing perch to another, the male does his utmost to drive away neighbouring rivals. The males of some species do not confine themselves to singing from a perch, but also have a song-flight that may be fairly brief, as with whitethroats and

serins, or long, as with larks. A few days later, it is the females' turn to make their appearance. From then on, the song has a dual function: to drive away rival males and to attract a mate. Once the female has been won, singers become less assiduous, and virtually silent when the young are being reared. Indeed, it is a safe bet that a male that continues to sing late into the season has remained single.

Song is not the only way to stake a claim to territory. In many species, posturing and ritualised behaviour play a role in drawing invisible boundaries. If these are crossed there is confrontation - usually brief and for show rather than violent, for intimidation is more important than actual violence. Water birds such as coots and grebes behave in this way, and raptors indulge in aerial attacks - shows of strength that are also meant to intimidate.

Among birds that nest colonially, gestures, accompanied by various vocal signals, are understood by all members of the group: the absent-minded or the foolhardy are soon called to order, sometimes harshly, under a hail of pecks. Gannets possess a repertoire of postures that are superbly choreographed; herons (such as grey herons or egrets) strike particular poses and raise their hackles; and gulls and terns partly open their wings and utter angry cries to frighten undesirables. In short, everything is organised so that all can enjoy their own home - if not in harmony with neighbours, then at least in mutual respect.

Left
Some Sardinian warblers barely leave their native region, while others must cross the Mediterranean after wintering in Africa. The males proclaim their territory by singing vigorously.

Right
The marsh-dwelling reed bunting (above) and bluethroat (below) must defend their small patches by singing relentlessly. The reed bunting's brief refrain sounds melancholy, while the bluethroat's varied utterances are playful and lively.

towards seventh heaven

Some species pair very early, in their wintering area.
Sometimes a pair has a lasting relationship and all that is
needed is to tighten conjugal ties as the breeding season
draws near. Common cranes and geese come into this category
of "faithful" birds. In other cases, however, the pair is formed
at the nesting site itself. Although a female may have been
attracted to a male's territory by a particular song or display,
the pair can only be formed if both future partners know the
code of behaviour applicable to that particular species, and
how to apply it. The precise ritualisation of mating displays
prevents, among other things, interbreeding between species,
but also acts as a channel for the substantial aggression that
exists between males and females of any given species.

In many species the male must have infinite patience and
great determination to overcome the female's reluctance to
allow any contact. His approach is gradual and cautious -
rebuffs and temporary failures are part of the game.
Eventually, reticence diminishes and the bond is formed,
but direct contact remains furtive.

Mating displays, discreet and measured or complex and
impressive, are always attractive to watch. They often involve
the swelling or exposure of part of the bird's plumage that is
striking in appearance or colour. Head decorations such as
crests or tufts are displayed to best advantage, as is the tail,
which may be raised, spread or fanned out. The gestures used
to woo a mate may include little jumps on the spot, bowing,
neck-stretching, spasmodic movements of some part of the
body or, by contrast, formal, motionless posturing. The wings
are often opened to display marks or coloured areas, and
agitated or vibrated. Among certain raptors and some waders
and ducks especially, displays may be aerial, consisting mainly
of chases and calls. Lapwings excel at this, and perform
stunning aerobatics, accompanied by mewing and sharp cries.
For many species, the display is not complete without the
presentation of gifts. Following a ritual developed in the
course of evolution, the male must offer his partner a
"present" in the form of prey or a morsel of food. The male
tern offers his female a small fish; the Montagu's harrier
brings his mate a vole, which she seizes in mid-air, performing
a graceful aerial arabesque.

Left
*With bill pointed skywards
in a slightly formal attitude,
a gannet scrupulously
performs one of the phases
of its species-specific ritual
display.*

Right
*This event must be seen -
and heard - to be appreciated
fully. The male Artic tern
beats his wings vigorously
while placing a small fish
into the gaping bill of the
female, the latter in begging
pose. This presentation is
accompanied by strident
calls.*

These gifts, which we should be careful not to see in human terms, are far from having a purely symbolic value: their purpose is to demonstrate to the female that the male will be able to feed her when she is confined to the nest while incubating, and be able to help feed the brood.

Often, though not necessarily, a display will end with mating, in which the female invites the male by adopting a particular position, usually a crouching posture, which signals her acceptance of him. This moment of climax demands the skills of an acrobat from the male, for he must keep his balance on the slippery back of his mate who, it must be said, helps by keeping absolutely still and sometimes by opening her wings slightly. The difficulties this causes for long-legged species such as herons, flamingoes, and storks can be imagined. Male raptors take the precaution of closing their claws, to avoid wounding the female. Swifts mate in flight, giving a new meaning to the phrase "in seventh heaven".

young male seeks love nest

The site chosen for the nest – or at any rate for the place where the eggs are laid, for we shall see that not all birds build a nest – is extraordinarily varied. Cliffs attract seabirds such as auks, gannets and kittiwakes, as well as raptors. In the last group, however, there are few migrants, perhaps because the best sites are reserved for resident birds which, occupying them all year, exercise a sort of monopoly over them. From the biggest boughs to the twigs at the very crown, trees are filled with all sorts of passerines, from rooks to thrushes, along with woodpigeons and turtle doves. Trees also attract raptor, herons and a few waders, such as green sandpipers. Bushes and shrubs, whether scattered or grouped together into hedges, are also home to large numbers of passerines: warblers, buntings, and blackbirds. Marshes, with their great flooded reedbeds, play host to birds such as water-loving warblers, purple herons, and marsh harriers. Stretches of water are frequented by grebes, gulls and marsh terns. Many birds select hollow trees or rock cavities. Hole-nesters include the migrant pied flycatcher, redstart, Scop's owl and, more surprisingly, ducks such as the goldeneye, goosander and merganser. The chicks of the latter have to leap from their refuge to reach the nearest water! Many birds nest on the ground, examples being most waders, terns and gulls, quails and passerines such as larks and some warblers. Finally, buildings have been adopted by a whole range of species, which use them as substitutes for cliffs. The best known include swallows, house martins and white storks, but also swifts, black redstarts, and raptors such as kestrels and peregrines.

Above
Puffins spend a great deal of time exchanging signs of affection. Rather than real grooming, these small gestures made with the bill symbolically express the strong attachment that exists during nesting.

Right
Among puffins (this one is returning from a fishing trip with a catch of sand-eels), bonds between partners are strong, and the same pair may meet up again year after year to nest.

Following pages
The closing stages of courtship between hoopoes. The female is insistently demanding the caterpillar that the male has brought her. He appears torn between the desire to satisfy her and a fear of approaching her. Mating displays are a way of gradually reducing the instinctive aggression between members of a pair.

The cuckoo's strategy

Some species have taken a radical approach to the problem of building a nest: they don't bother at all. These birds are content to lay their eggs on the bare rock of a cliff, as do guillemots, or directly on the ground, often in a small hollow, as do Arctic skuas and pratincoles. Waders such as plovers, lapwings and oystercatchers are content to gather a few fragments of shells or shreds of vegetation in a small hollow they have scraped with their breast in the earth or sand. The female eider duck uses her own down as the sole lining of the hollow that serves as her nest. Some birds simply use the old nest of another species. The hobby, on returning from Africa, takes over an old crow's or rook's nest. The cuckoo and great spotted cuckoo go one better, for not only do they not build a nest, but they lay their eggs in those of other species: small passerines in the cuckoo's case, and magpies in the case of the great spotted cuckoo.

However, building a nest is *de rigueur* for many, especially among passerines and water birds. Migrants, even more than other birds, are fast nest builders. There is no time to lose, so the job is done in the matter of a few days. In this respect white storks and ospreys show initiative by refurbishing last year's nest with a few extra branches. As each refurbishment is added, the edifice can reach an impressive size, sometimes to the point of bringing down the supporting tree.

Luxury or spartanism

For some species, ranging from European bee-eaters to sand martins, shearwaters and puffins, the nest is no more than a burrow. Other birds - the true builders - use all sorts of materials: blades of grass, stems, straw, twigs and branches are all widely used. To shape their nest into a cup - which is more or less closed depending on the species - swallows use mud which, once dry, forms a very strong structure.

Nest builders display a variety of talents. The turtle dove's nest is a simple platform made of twigs; that of the reed warbler is a deep bowl of grass and stems skilfully attached to a few reed stalks. The penduline tit's nest is a true architectural masterpiece: a sort of soft, padded bag, entered from the side and suspended at the end of a twig - usually over water to increase safety for the brood.

The materials used for lining the nest are chosen for their softness and often for their insulating properties, which help to keep the eggs warm. These are often feathers, hair or fibres of vegetable origin, as well as man-made materials such as wisps of wool, all kinds of string and bits of paper or plastic. Sometimes, as indicated above, the nest's lining is a mere formality: a few straws, stones, five or six pebbles or a few rabbit droppings are enough to provide a rather spartan level of comfort.

Above
By receiving an insect as a mating gift, the female hoopoe can gauge the male's ability to feed her when she is confined to the nest.

Right
When displaying, the wall-creeper spreads its broad wings, adorned with carmine patches and white spots, which make this small mountain bird with its agile flight look like a large, brightly-coloured moth.

ONCE THE PLACE FOR EGG-LAYING has been chosen, and the nest – if there is to be one – built, a miracle happens, as it does every spring. If all goes well, within a few weeks the young birds will appear.

Happiness is a nest

For their parents, this marks the start of the busiest period of the year. There is not a second's respite, whether they are feeding or simply watching over their young – depending on whether these remain in the nest or, as we shall see, leave it soon after hatching. Migrants, even more than other birds, must hurry to finish breeding before the unstoppable march of the seasons imposes itself and they again feel the need to head south.

Top left
Protected from the cold by their thick down, eider chicks cling to their mother for reassurance.

Right
The red-throated diver must wait almost a month before its eggs hatch and allow it to leave the islet where it has built its nest, out of reach of land predators.

eggs under maximum security

Between April and June, hidden in bushes, in the tangles of branches, deep in the marshes, and on beaches and cliffs, billions of birds' eggs are laid, all over Europe. This is impressive in itself, but just as astonishing is that such a vast process takes place almost unnoticed. Of course, this secrecy is an excellent safety measure. Birds that do not nest secretively are those that breed colonially, as do many seabirds and most herons. Their strategy relies on a common interest: the large number in their group generally assures that each member is safe. If a predator, such as a crow or fox, shows interest in a colony of gulls or terns, the latter close ranks and take turns to attack the intruder relentlessly. Deafened by strident cries and harried by pecks, the intruder soon feels threatened and usually retreats without further enquiry! The little black-necked grebe fully understands how it can benefit from this constant

watchfulness and builds its nest among those of black-headed gulls, thus enjoying effective protection without having to contribute.

By contrast, for non-colonial nesters, such as almost all passerines, it is essential not to be spotted or attract the attention of prowling predators to the nest. Their eggs are, therefore, laid well out of sight – later we shall see how this caution continues once the nestlings have hatched.

warm atmosphere

For an embryo to develop inside an egg, it must be kept at a certain warm temperature for a given period of time. This is called incubation and is often the task of the female, which has "incubation patches" on her belly. These are patches of bare skin, whose subcutaneous blood supply is increased at nesting time to provide the warmth the eggs need. When she sits on the eggs, she parts her belly feathers in order to place the incubation patches directly against the eggs. If the female is the only one to incubate the eggs, the male may only prevent the clutch from cooling when the female is forced to leave the nest. In some species, however, the male and female have the same physical features and the pair share incubation more or less equally. This equality can be seen in the starling, roller, house martin, and little ringed plover. When the female alone incubates, as in the red-backed shrike and hobby, it is not unusual for her partner to provide almost all her food, often

Above (left to right)
A short history of the black-necked grebe: courtship, mating, incubation and hatching. A present from the male to the female seals their union; mating takes place under the gaze of a pair of whiskered terns; the two birds share incubation; and a chick rides comfortably on its parent's back across its native pool.

Right
This photograph cannot convey the hubbub of a colony of black-headed gulls. In constant movement, each bird expresses itself forcefully, as if trying to drown the cries of its neighbours. The results are deafening.

bringing it directly to the nest. Incubation does not take long. Among passerines it takes just a few weeks: 11 days may be enough for the skylark and siskin, and a week or two longer for most of the bigger species. Thus incubation takes about three weeks in species as widely different as the swift, woodcock, black tern and garganey, and just under a month for the common scoter, white-fronted goose and lapwing. As a general rule incubation is short for most migrant species, which are almost always breeding under a time constraint. Even a species as large as the whooper swan (which can weigh as much as 22 pounds, and whose eggs may weigh 10 ounces - compare the $\frac{1}{28}$ ounce weight of a willow warbler's egg!) only incubates for just over a month. Because of these same time pressures, many migrant birds - especially those that travel long distances - raise just one brood per year, whereas many resident birds and partial migrants have time to raise two or more. But, as always in nature, there are exceptions. Swallows, which are indefatigable migrants, may well raise two or three broods between April and October.

If there is a problem, for example, if the clutch is destroyed by a predator or by a long period of bad weather such as persistent rain or bitter cold, or if the young die or are caught by predators while still small, a second clutch may be laid. This is known as a "replacement clutch". For reasons outlined above,

migrant birds can only do this if enough time remains. Little egrets, for example, only make a second attempt if a problem occurs at the beginning of the breeding season. The white pelican is even more rigorous: it only lays a replacement clutch if the first was lost fewer than 10 days after laying. For all these birds, if the nest is robbed or destroyed at too late a stage, replacement of the clutch or brood would take too long, so the adults are forced to abandon nesting for that year. And some species, such as the pink-footed goose and fulmar, never lay a replacement clutch, even if there might appear to be enough time.

NO sooner hatched ...

Not all newly-hatched chicks are the same. They can be divided into two groups: nidicolous (nest-dwelling) chicks - which are helpless at birth, usually naked, with their eyes closed, and are totally dependent on their parents for warmth and food - and nidifugous (nest-fleeing) chicks, born covered in down and with their eyes open, which can move about and feed almost at once.

Left
Stoical under a downpour, a female short-toed eagle sits on her eggs.

Right
Hidden behind a protective screen of reeds, a female great crested grebe incubates her clutch while waiting for the male to relieve her, which may take several hours.

The former need only a fairly short incubation period, but need extra time after hatching before they can achieve a degree of independence. The incubation period needed by the latter is longer, but they are relatively independent almost as soon as they are hatched. The chicks of passerines, woodpeckers, diurnal and nocturnal raptors, and herons are nidicolous; those of waders, quails, partridges, ducks and grebes are nidifugous.

Depending on the species, both parents may look after the brood - as with grebes, geese, cranes and raptors - or only one, usually the female.

Sometimes the male confines himself to keeping watch over the young while the female looks after them more closely, especially by feeding them. Among raptors, the male brings food to the nest, but it is the female that tears it up and distributes it among the chicks. Male and female passerines share the task of feeding their brood roughly equally, but if the female lays a second clutch it is the male's job to oversee the first brood for a few days, until they can fend for themselves. Among many species of duck, the female alone looks after the ducklings, the male's sole contribution having been to mate with her. By contrast, among waders nesting in Arctic regions, the male is often responsible for the chicks: as soon as the female has laid and finished incubating, she flies south once again. Exhausted by the combination of a long migration from Africa to the Arctic tundra and the production of eggs, she passes the baton to her partner, who is in better physical condition. Many of the waders present in central Europe in summer are, therefore, females on their way back to their winter quarters.

The art of concealment

In species whose chicks are nidicolous, caution governs the behaviour of the parents while feeding them. Just as the nest was carefully hidden, so the endless to-ing and fro-ing with food for their insatiable brood takes place discreetly.

The parents only approach the nest after checking that no danger threatens it. For the same reason, they often do not go there directly, but will first perch on a nearby bush and approach in stages, from twig to twig, or they will alight on the ground a few feet away from their brood and reach the nest by passing hidden through long grass. If a bird about to return to the nest spots danger it will remain at a distance for a long time, often making a constant alarm call. If the intruder does not move away, the bird may eventually swallow the food intended for its chicks and leave temporarily, thus interrupting feeding for a time.

Exceptions to the rule of extreme caution are to be found (aside from species that nest colonially, as already described) among birds that have chosen to live close to people. Thus swallows and storks go about their various nesting activities without making the slightest attempt to hide them. This confidence can be explained by the great warmth humans feel for them: as symbols of happiness, their return is eagerly awaited every spring.

Above
The male (top) and female stonechat share the arduous task of feeding their young. Over two weeks, thousands of insects will be deposited in the chicks' eager bills.

Right
Whiskered terns at the nest. While one of the adults (top) takes charge of feeding, the other stays close to the very young chicks. One is hidden under its parent's belly while the other, steady on its little legs, looks round inquisitively.

Because of this need to remain unobtrusive, evolution has provided nidicolous chicks with a special system for dealing with droppings. A nest stained with white excrement would soon become highly visible and attract the attention of predators. However, if the chicks' droppings had the same runny or liquid consistency as those of their parents, the latter could not pick them up easily. But nature has got round this snag, like so many others: the chicks' droppings are enveloped in a gelatinous substance, which makes them easy to carry. The adults can thus carry these "faecal sacs" in their bills, and drop them far from the nest. The problem is often solved even more directly, at least in the early stage of rearing young, by the parents simply swallowing the brood's droppings. While this might seem unsavoury, it works to the birds' advantage because of the nutrients the droppings contain.

Freedom for the young

Among nidicolous birds the time spent in the nest varies, like incubation, in proportion to their size. But the ability to leave the nest does not depend on the power of flight. The young birds often leave the nest by fluttering awkwardly, even by walking or hopping. The chicks of small passerines only stay in the nest for about two weeks; some, such as those

of the reed warbler and woodlark, can leave after barely 10 days. But the chicks must wait for their feathers to finish growing, and then learn to fend for themselves. While starlings are left to make their own way a few days after leaving the nest, young garden warblers take several weeks, and woodchat shrikes a month and a half, to become independent. At this point, the young birds can start to disperse - a process described in the next chapter.

Juvenile plumage

Whether they are nidicolous or nidifugous, all chicks go through a first stage when their "plumage" consists of very short, loose feathers - down. Among the former this is often white or grey. By contrast, in chicks that leave the nest early, down needs to offer camouflage protection.

Left
With one baby on her back and the rest sheltering under her broad wings, this female mute swan is completely taken over by her brood. Swans are known for being strongly attached to their young.

Right
All the Arctic tern's grace and lightness are captured in this wonderful shot of an adult feeding its partly-grown youngster.

Following pages
The fulmar is not in the habit of removing eggshells from the nest after hatching. It is hard to imagine that this chick, whose down is barely dry, will, within a month and a half, be able to fly with all the grace of its species.

It is, therefore, the colour of moss, bark, lichen or stones, mixing brown, russet, black or grey, in complex patterns. This down is gradually replaced by the young bird's first feathers, known as juvenile plumage, which is often a rather drab copy of the adult's. The hues are less striking, marks and patterns subdued, and sheen dulled or absent. This can be seen in species such as geese, bee-eaters, turtle doves, pigeons and swallows. In other species, the juvenile plumage is strongly reminiscent of that of the adult female. Ducks, notably, as well as harriers, certain falcons and passerines such as redstarts and buntings come into this category. Finally, in some birds the difference between adults and juveniles is such that it is hard to believe that two individuals of different ages belong to the same species. An adult night heron (a small, nocturnal heron) with its handsome grey, black and white plumage looks quite different from a juvenile, greyish-brown with white spots, quite apart from its shape. A gulf separates the adult starling, which is black with sheens of various colours, from the juvenile, which is entirely greyish brown.

Among the smaller species, which include the majority of passerines, juvenile plumage is shed fairly quickly - by autumn or, at the latest, by the winter. The young bird then displays new feathers, known as "first winter plumage" and, from then on, it becomes quite hard to tell older and young birds apart. Many medium-sized species retain some of the juvenile plumage, which allows identification of the young birds, though the differences may be subtle: this can be seen in terns, gulls and certain waders. The last category is made up of larger birds, whose size renders their development slower. Large raptors such as eagles and vultures, as well as large gulls, go through various stages of plumage over several years before they gain the full adult pattern.

It is in their first winter plumage, which grows progressively, that many young birds embark on their first autumn voyage; indeed, juvenile plumage has the advantage that it grows quickly, replacing down, and allowing the young birds to begin their journey as quickly as possible. However, this accelerated growth has the disadvantage that it produces feathers of relatively poor quality, which are therefore weaker. Migrants, therefore, must equip themselves, through at least a partial moult, with better quality feathers before undertaking migration.

Above
Straining with its legs and head, a cuckoo chick that has only just hatched ejects an egg laid by the warbler whose nest it occupies. Instinct drives it to do this so that it remains alone in the nest, and is the sole beneficiary of the food brought by its adoptive parents.

Above right
Huddled in the undergrowth, these snipe chicks are protected by their plumage, whose complex patterns blend with the surrounding moss and leaves.

Below right
Having left the nest before it can even fly, this young skylark must wait several more days to let its wing feathers develop and allow it to make its maiden flight.

<u>After their pampered time in the nest</u> and their discovery of the outside world with their parents, young birds become independent.

The young take flight

At first they do not seem to know what to do with their new-found freedom, all the more so since they are often driven away by their parents who, having accomplished their mission, soon lose patience with their offspring. This is a slightly uncertain time of scattering, which immediately follows fledging, and is a prelude to their true departure on migration. Torn between their need to rest and the desire to leave, some adults also experience this confused time of dispersion.

Top left
Comfortably ensconced in the clay bowl that serves as their nest, these young swallows quietly await the return of their parents, bearing insects caught on the wing. After moulting, their ochre-coloured throats will eventually become brick-red like those of their parents.

Right
Already as big as its parents, this young lapwing is now independent and can, literally and metaphorically, stand on its own two feet. Nevertheless, certain signs distinguish it from an adult: its crest is very short, the sides of its head are not yet white, the dark marks around its eyes are not yet clearly defined, and the feathers of its upper parts have pale edges.

Into the unknown ...

For young birds of species that retain family links at least until the beginning of winter, such as swans and geese, migratory departure is relatively straightforward. They can be guided and follow other birds without having to adapt to doing things independently. Despite this, individuals may stray from the group and find themselves alone and more or less lost. For all other species, among which family ties are ephemeral, the young birds are left to fend for themselves, relying on instinct. Many of these birds disperse for a few weeks before deciding to set off for their wintering areas. Although there is a general southward drift, this postnuptial dispersion sometimes leads members of the next generation to places where they would not normally go. However, by following the example of another member of their species or driven by weather conditions, these birds tempted to explore new places usually return gradually to the right route. Occasionally they may delay doing so and, in the event of sudden bad weather, meet a tragic fate.

The time of wandering

The vast numbers of young birds - far more numerous than adults at the end of summer - mean that they are everywhere. Hedges, bushes and undergrowth rustle with the sound of young wings being exercised. On pools, in marshes and along coastlines, the duller, more subdued plumage of numerous juveniles predominates. Young passerines flutter about, some still uncertainly. Young ducks, whose plumage resembles that of adult females, do not yet know what dangers they will soon face. Young raptors continue learning to hunt, a process begun under the supervision of their parents. There is no question, yet, of beginning their long journey. Summer is still here, even though there are signs that it is fading. Storms, coolness at night and, in northern Europe, the first snows, all signal that autumn is well and truly on the way.

Meanwhile, many young migrants move about as the mood takes them, depending on the opportunities that present themselves on their temporary wanderings. Others linger in welcoming spots where safety and food appear to be assured. Although they can probably discover such places for themselves, no doubt they follow the example of older birds, which also seek out such sites. However, the fact that young and older birds of the same species frequent the same favourable places in large numbers does not mean they are moving according to the same strategy: it is more that the quality of these sites means they coincide there. But, whereas the adults are driven by a definite impulse to migrate and already have some idea of their route and the destination they are aiming for, young birds have more of a tendency to take a break, to "idle" somewhat. However, this interlude soon ends. Migration is about to begin in earnest, and all these young birds must actively prepare for it.

> "sometimes a great bird casts its shadow, a long-legged hermit of the pools that lie all around."

Frédéric Mistral

Above
A female pochard bathes without taking her eyes off one of her ducklings. Not all individuals nest at the same time across that species' range. Among pochards, for example, ducklings can be seen from mid-May to the end of July. Dates of shooting seasons must take these variations into account.

Right
This young grey heron, in its wandering phase after leaving the nest, does not yet show the handsome plumage of the adult: it lacks the long feathers on the back of the neck of sexually mature birds.

whether born during the spring or belonging to an older generation, migrant birds must now think of setting off for their wintering areas once more.

departure

Some - as we saw in the second part of this book - will travel only a short distance. Others will, for the first time, cover thousands upon thousands of miles on a voyage that will take them across farmland, mountains, deserts and seas.

Top left
For white storks, the start of their long voyage is drawing near. After spending several months bringing up their young in Denmark, France or Spain, they will set off once again for the African savanna. However, in recent years some have taken to staying in Europe for the winter, notably in southern Spain.

Right
Soon this whimbrel - seen perching in a tree as some waders do at nesting time - will be pattering about the mudflats of Europe's Atlantic coast, on its way south.

swaLLows and romanticism

The idea that bird migration only takes place in the autumn comes from romantic imagery. Poets are largely responsible for creating this stereotype by frequently associating autumn with the departure of symbolic birds such as swallows. Similarly, it is often thought that all migratory birds leave at more or less the same time, as if obeying some mysterious signal. Such beliefs are erroneous, however. In fact, postnuptial migration begins as early as July, especially in northern Europe and in Arctic latitudes. In these regions summer lasts only two or three months; birds breed as soon as they arrive there, and leave as soon as nesting is over – a stay lasting a few weeks. This promptness is characteristic of divers,

many waders, northern ducks and passerines such as snow and Lapland buntings.

If we subtract the time needed for pairing, nest-building, incubation, and the growth of a young bird's feathers, we can see that young birds of certain species have only about three weeks in which to prepare for migration. Once this period has passed, deteriorating weather conditions would jeopardise their survival.

Even in central Europe, certain groups of birds hurry to reproduce before the weather imposes harsh constraints. Swifts, for example, arrive in our latitudes only towards the end of April, and most leave from the end of July through to early August. After this time, even though there may be a few stragglers, there is no more tearing around at roof height in the towns and villages of Europe, shrieking in the warm twilight. The brevity of its stay – a mere three or four months – has led the swift (and other birds such as the cuckoo, bee-eater, nightjar and various passerines) to be considered "African", because it spends two or three times as long in Africa as in Europe!

Migration departure dates therefore vary widely, and depend on geography as well as the behaviour of each individual species. Partial migrants probably show the greatest tendency to leave at different times. Rather than being driven by an internal mechanism, these birds are encouraged to move by local conditions, sometimes late in the season.

Above
Little time remains for the denizens of Spitzbergen to perpetuate their species. In a few weeks, snow and ice will have reclaimed the shores of King's Bay.

Right
By their calls, these two adult Arctic terns seem to be encouraging their fully-grown youngster to take flight. All three will soon travel from the northern icefields to the shores of the Antarctic, at the opposite pole.

Bingeing on sugar

As we saw earlier, the birds' departure on migration and the state of their plumage are closely connected in one way or another. Many migratory species set off after moulting has equipped them with new feathers, or at least with fresh primaries. Those that migrate in autumn before renewing their plumage have often done so in late spring, so that their feathers are still in good enough condition to last until the winter before being replaced. However, while plumage in good condition is essential for a migrant, the bird must also be capable of the intense and sustained effort required for migration, especially its longer stages. In other words, its wing muscles must be able to function at their physiological best, sometimes for long periods. The bird needs "fuel" to "keep its engine running". This energy source is simply the fat that when burned by the body makes muscular activity possible, and provides far better performance than carbohydrate or protein. It is simply a matter of storing enough of it. To do this, as migration time approaches birds generally indulge in what scientists call "hyperphagia" – a sort of bulimia. They need to eat a great deal in order to get in condition for their journey by building up this much-needed fat. Their bodies produce it largely from the sugar contained in the food they eat.

Many migrants therefore go looking for sugar-rich food. During this time of preparation for the autumn departure certain passerines – for example warblers, flycatchers and thrushes – eat large amounts of fruit or berries high in sugar. Some normally insect-eating birds even modify their diet at this time by becoming fruit eaters. Elderberries, redcurrants and wild cherries are consumed in nesting areas or at stopping points further south, such as around the Mediterranean. Fig trees, too, are enthusiastically frequented by migrating passerines such as orioles and flycatchers, the latter being known locally by the evocative name of "fig-peckers".

Amazing travellers

Whatever they have eaten, the migrants' bodies lay down substantial fat reserves, to the point that a short- or medium-range migrant's body weight increases by between a tenth and a quarter, while a long-range migrant's increases by between a third and a half. A warbler that normally weighs $^2/_3$ ounce or so manages, by eating large amounts, to store some $^1/_3$ ounce of fat. In all birds such stores are distributed around the body, in organ tissues, and under the skin where they form fatty deposits in places like the sides of the breast, the belly, the rump and in front of the breast muscles – which themselves may increase in mass by a third before the bird begins migration.

Above
At the end of winter, before spring migration – or at the end of summer, before the autumn departure – berries and small fruit are a precious resource for many migrants, such as these blackcaps (female above), gorging on ivy berries.

Right
This blackbird, gulping down a rowan berry, is building up its reserves before setting off.

Following pages
At migration time birds may stop off in unexpected places. This red-backed shrike, photographed amid sunflowers, would never normally be found in such a large cultivated area.

The extra weight of fat is partly offset by a reduction in the body's levels of protein, carbohydrate and water. Another remarkable physiological characteristic of migratory birds is the way they meet their need for large quantities of water to replace that lost through muscular activity: the water they use is actually produced by the combustion of fat to fuel muscular activity. For migrants, therefore, fat is a sort of miracle substance - not only is it an excellent fuel, but its very combustion provides the water without which they cannot function.

Since fat is like a fuel, it has been possible to calculate how far migrants can travel. All that is needed is to know how much  energy it takes to use an ounce of body mass during an hour of flight, and the calorie content of an ounce of fat. Thus it has been calculated that with $\frac{1}{3}$ ounce of fat a passerine could in theory fly for 30 hours at an average speed of 19 miles per hour - which means it could travel 560 miles without exhausting its fat reserves! Among stronger-flying migrants, such as ducks or large waders, these theoretical figures are astonishing, of the order of several thousand miles. The real figures are usually somewhat lower, even though the most powerful birds, especially swans and geese, sometimes come close to them in favourable conditions - that is, with a following wind.

Free as the wind

All this makes it easier to understand just how much migratory birds need ready supplies of food and places where they can feed safely. In Europe, rampant urban development due to demographic pressures or tourism, rapid changes in agriculture with the use of pesticides and habitat destruction, and the shrinking of wetlands, all help to destroy what migratory birds desperately need. A migrant that cannot replenish its fat reserves in time is doomed. It is vital, therefore, that these astonishing travellers be saved as far as possible, while there is still time. The existence of enough reserves - both in number and in surface area - should be complemented by global policies for nature conservation and wildlife protection.

How could we contemplate the disappearance of the magical sight of these birds, speeding twice-yearly in their endless search for some remote, ideal place? Driven by unstoppable, vital urges, the most indefatigable of these travellers carry mankind's dreams of freedom from one end of the earth to the other. We can only watch in wonder as they trace across the heavens their overwhelming desire for eternal summer.

Left
The island of Hoëdic, like others in the Atlantic, often plays host to vagrants. This Radde's warbler is thousands of miles from its native Far East.

Right
Highlighted against a leaden sky, common cranes fly over a Spanish landscape once more, as they have done for thousands of autumns.

practical hints

the well-equipped amateur ornithologist

Binoculars: 7× or 8× magnification, with objective lens diameter of at least 30mm.
Telescope (long-range): 15× to 30× magnification, with wide eyepiece (code W).

how to maximise your chances of success

Visit a reserve, such as one of the sites mentioned opposite – where advice will usually be available – or pick a place of your own, making sure you position yourself on a vantage point that gives an uninterrupted view, with the sun behind you.

Choose a dry day, preferably with a light wind in the same general direction as migration: in spring, from the south; in autumn, from the north. In autumn, an area of low pressure over northern Europe will increase the chances of seeing birds pass over.

At regular intervals, scan the sky with your binoculars in a slow sweeping movement from left to right, or vice versa, holding them at an angle of between 30° and 45°. Listen for the calls of migrating birds: you often hear them before you see them.

If sea-watching, sweep your binoculars across the horizon, keeping about two-thirds sea and one-third sky in the field of view. Make sure your movement is in the opposite direction to that of the migrants, as this makes it easier to spot birds on the move.

watching birds on passage

You can watch migration in many places, but there are certain sites where birds congregate, and these are the best for observing "active" diurnal migration – that is, birds on passage proper. A selection of the most outstanding sites is listed here, along with some notes about their characteristics.

Migrants can also be seen when they break their journey, when they are wintering or at their breeding grounds. A selection of the best sites is listed. These are mostly protected sites, which means you can watch the birds in peace, and fully equipped hides are often provided.

Above, from left to right
Common crane, roller,
black-headed gull,
black-necked grebe,
bearded tit, nightingale,
spotted redshank, shag.

selected european bird reserves or migration hotspots

Each of the following sites is worth a journey in itself, and needs a stay of several days to be fully appreciated. All tend to be at their most interesting in spring and autumn.

UK and Ireland

¬ Cley
Location: north Norfolk coast.
Season: winter, spring, summer, autumn.
Species: seabirds, waders, wildfowl, migrant passerines.
Notes: one of Britain's top reserves, and a magnet for rare species.

¬ Isles of Scilly
Location: south-west of Cornwall.
Season: spring, autumn.
Species: seabirds, migrant passerines.
Notes: often has American vagrants.

¬ Cape Clear Island
Location: extreme south-west of Ireland.
Season: spring, summer, autumn.
Species: seabirds, migrant passerines.
Notes: top site for passage seabirds, especially shearwaters; annual American vagrants.

¬ Wexford Slobs
Location: near south-east tip of Ireland.
Season: winter, autumn.
Species: wildfowl, waders.
Notes: best known for its large numbers of Greenland white-fronted geese, pink-footed geese and occasional snow geese; also Bewick's swans, sea ducks and waders.

some other European sites

Austria

¬ Neusiedlersee
Location: eastern Austria, near Hungarian border.
Season: spring, summer.
Species: herons, raptors, warblers.
Notes: splendid shallow steppe-lake with large colonies of breeding herons and other birds.

Denmark

¬ Waddensee
Location: North Sea coast.
Season: winter, summer, autumn.
Species: waders, wildfowl, seabirds.
Notes: Europe's most important intertidal habitat, with huge gatherings of waders.

France

¬ Cap Gris-Nez
Département: Pas-de-Calais.
Location: coastal.
Season: autumn.
Species: especially seabirds: terns, skuas, auks, divers, scoters, eiders. Also passerines and raptors.
Notes: choose a time when wind is chiefly from the north or north-west. No visitor centre.

¬ Lac du Der-Chantecoq
Département: Marne.
Species: common cranes (passage migrants and wintering birds); raptors: white-tailed eagles, kites, peregrine falcons, merlins; ducks, geese.

¬ Organbidexka
Département: Pyrénées-Atlantiques.
Location: mountain pass of medium height, between Iraty and Larrau.
Season: summer, autumn.
Species: raptors: kites, honey buzzards, ospreys; storks, cranes, pigeons; passerines: swallows, wagtails and others.
Notes: "Organbidexka col libre – pertuis pyrénéens" (OCL) reception and information centre open from 15 July to 15 November (05 59 04 87 50). There is also a birdwatching site at the Lizarietta Pass, south of Sare.

Greece

¬ Evros Delta

Location: forms border between Greece and Turkey.

Season: winter, spring, summer, autumn.

Species: herons, waders, wildfowl, pelicans, raptors, warblers.

Notes: outstanding wetland, with extensive lagoons and reedbeds; wintering wildfowl include lesser white-fronted and red-breasted geese.

Netherlands

¬ Texel

Location: island off the Dutch coast.

Season: winter, spring, summer, autumn.

Species: waders (notably black-tailed godwits and ruffs), spoonbills, harriers.

Notes: largest of the Friesian islands, with good dunes, mudflats and saltmarshes.

Sweden

¬ Falsterbo

Location: coastal (promontory south of Malmö).

Season: especially autumn.

Species: passerines: finches; raptors: sparrowhawks, buzzards and rough-legged buzzards, golden eagles.

Notes: a large proportion of Scandinavia's migrants gather here on the way to their next stopping point, Denmark.

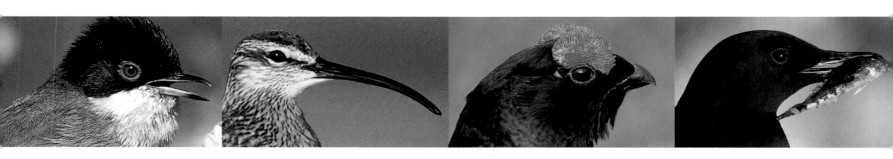

Hungary

¬ Hortobagy

Location: eastern Hungary.

Season: spring, summer, autumn.

Species: splendid raptors including red-footed falcons, white-tailed eagles, peregrines and saker falcons; breeding great bustards, and lesser white-fronted geese and cranes on passage.

Notes: steppe habitat with woodland, wetland and pasture.

Israel

¬ Eilat

Location: coastal.

Season: autumn, spring.

Species: raptors: eagles, kites, sparrowhawks, long-legged buzzards; storks, pelicans; passerines.

Notes: one of the northern hemisphere's most important migration sites. Vast numbers and a huge range of species, including some from western Asia.

Portugal

¬ Cape St Vincent

Location: south-west tip of Portugal.

Season: winter, spring, summer, autumn.

Species: seabirds, migrant passerines, choughs, peregrines.

Notes: fine vantage point for observing passages of seabirds, notably shearwaters and petrels.

Spain

¬ Straits of Gibraltar

Location: coastal (Tarifa headland and around).

Season: autumn, spring.

Species: raptors; storks; passerines; also seabirds.

Notes: one of Europe's most important migration sites. Wide range of species, often passing in huge numbers.

Turkey

¬ Bosporus

Location: coastal (Istanbul).

Season: autumn, spring.

Species: raptors; storks; passerines.

Notes: one of Europe's prime migration sites. Vast numbers and wide range of species.

Bibliography

Most of the following titles have become classics. They offer the reader comprehensive information on migratory birds.

T. Alerstam, *Bird Migration*, Cambridge University Press, 1993.

P. Berthold, *Bird Migration: A General Survey*, Oxford University Press, 1993.

J. Dorst, *Les Migrations des Oiseaux*, Payot, 1956.

P. Géroudet, *La Vie Des oiseaux*, Delachaux et Niestlé; seven volumes on passerines, raptors, waders, etc. Several editions from 1951 to 1998.

acknowledgements

The author would like to thanks all those who, in various ways, helped in the production of this book or in searching for documents, especially:

J. Chevallier, A, Coulon, A. Czajkowski, L. Dard, M. Duquet, B. Kosel, Ph. Martin, S. Nicolle, C. G. Petrow, É. Piéchaud, P. Olombel, L. Pierrot-Pradel for research and translation of quotations from Spanish literature, as well as C. N. Vermeer and I. Kharitinova.

The invaluable help of Odile Perrard allowed much-needed improvements to the text.

The design of Anne-Marie Bourgeois, and her work on making the best use of photographs, have breathed a poetry into this book that make it an even more glowing tribute to the beauty of these birds.

Finally, part of this book is the work of Yves Calvi, to whom I give my warmest thanks.

Above, from left to right
Sardinian warbler, whimbrel, black grouse, black guillemot, Lapland bunting, goldfinch, turnstone, dunlin.

Following pages
Common cranes against the moon.